ANIMAL
AND PLANT
MIMICRY

ANIMAL AND PLANT MIMICRY

Dorothy Hinshaw Patent

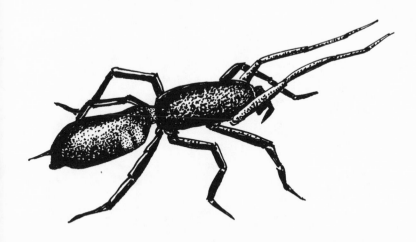

HOLIDAY HOUSE · NEW YORK

For Mom and Dad, who always encouraged
my interest in living things

Library of Congress Cataloging in Publication Data
Patent, Dorothy Hinshaw.
  Animal and plant mimicry.

  Bibliography: p. 121
  Includes index.
  SUMMARY: Discusses examples and possible
causes of plant and animal mimicry.
    1.   Mimicry (Biology)—Juvenile literature.
[1.   Camouflage (Biology)]   I.   Title.
QH546.P35   574.5   78-7457
ISBN 0-8234-0331-9

Contents

1 • How To "Fool" Your Enemies · · · 7

2 • Saga of the Monarch and the Viceroy · · · 20

3 • Beauty on the Wing: Tropical Butterflies · · · 30

4 • How To "Fool" Your Food · · · 44

5 • Mimicry in "Cold-Blooded" Animals · · · 62

6 • Mimicry of Social Insects · · · 79

7 • Plants Do It Too · · · 96

8 • Puzzles, Problems, and Proof · · · 105

Glossary · · · 119

Suggested Reading · · · 121

Index · · · 123

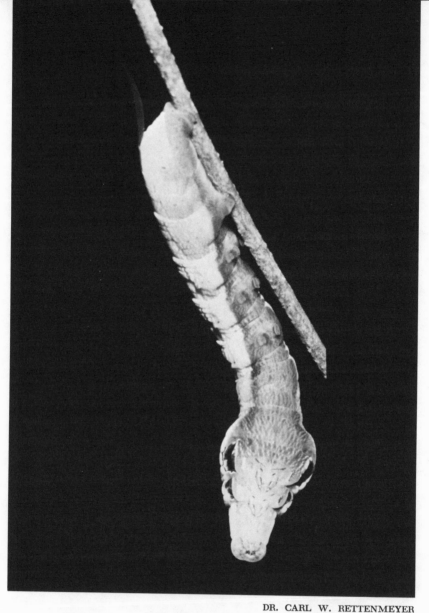

DR. CARL W. RETTENMEYER

This is not a snake but the caterpillar of a sphinx moth found in Ecuador. When disturbed, it twists its body and expands its thorax to reveal a pair of large false eyes—actually spots— that make it resemble a poisonous viper.

1 · How To "Fool" Your Enemies

Have you ever seen a viceroy butterfly and mistaken it for the more familiar monarch? Or have you seen what you thought was a bee hovering near some flowers and then realized that the creature was really a yellow fly with black stripes? Throughout the living world there are animals, often quite unrelated, which may look so much alike that even expert scientists are fooled.

When one living thing resembles another so closely that it is in some way mistaken for the other, it is called a mimic. What it resembles is called its model. The phenomenon of such similarity between species is called mimicry. While most people are unaware of it, mimicry abounds in nature, especially among insects. Butterflies tend to mimic other bad-tasting butterflies, while stinging bees, wasps, and hornets are copied by a great variety of other insects such as bugs, moths, beetles, and flies. Ants are mimicked by many beetles and spiders, while grasshoppers and crickets may resemble bombardier beetles and tiger beetles. Amazing examples of mimicry are known among birds and fishes, and mimicry is suspected

7

in some mammals. As we shall see, even some plants mimic one another, while others "copy" insects.

In discussing mimicry, it is almost impossible to avoid using words such as "copy," "fool," and "deceive." These words usually imply a conscious intent on the part of the "copier," but not in the case of mimicry. Animals and plants do not have a mentality which allows them consciously to copy another living thing. If a tasty butterfly looks and behaves like a distasteful one, it has no awareness of the fact. It merely looks and behaves the way its heredity determines. The shape, size, and color of its wings are produced by the interaction of many inherited traits, while its way of flight and perching are determined by the shape and size of the muscles in its body, not by any conscious attempt to imitate another kind of butterfly. Remember that whenever such words as "disguise" or "copy" are used in discussing mimicry, they are not meant to imply any conscious effort on the part of the "deceiver."

Early Studies

While people noticed long ago that certain flies looked very much like bees, no one really thought much about the phenomenon of mimicry until the late nineteenth century. The first scientist to sit down and try to figure out why such resemblances should occur was an English naturalist and explorer named Henry Walter Bates. This investigator had spent eleven years in the jungles of the Amazon, where the variety of insect life was astounding. Bates especially enjoyed studying the brightly colored tropical

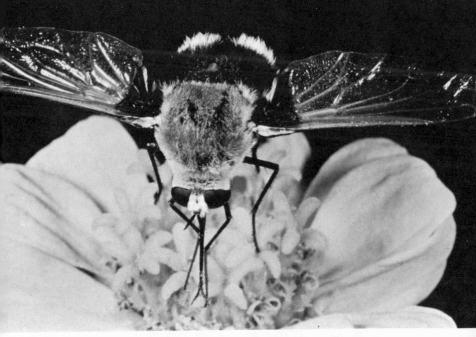

DR. CARL W. RETTENMEYER

It's quite remarkable how much this bee fly from Tanzania resembles a bee.

butterflies. He was puzzled to see very abundant, bright-hued butterflies flying slowly and fearlessly in areas where insect-eating birds were plentiful. When he looked more closely at these pretty creatures, he was surprised to find butterflies which looked almost identical but which were from completely different families flying together. The most common butterflies were from a tropical family called the Heliconidae, or passionflower butterflies. But along with the heliconids he found members of a very different family, the Pieridae. Most pierids, such as the common cabbage butterfly, are white or yellow. Because of these colors, pierids often have names like whites, sulphurs, and brimstones. But these Brazilian species were very different. They were almost identical in color and pattern to the beautiful black, yellow, and red heliconids among which they flew. Their wings even had the more elongated shape characteristic of heliconids instead of the typically rounded pierid shape.

After much thinking, Bates reasoned that the heliconids must be bad-tasting butterflies. Even though they flew slowly and their bright colors made them obvious, birds seemed to ignore them. Bates decided that by resembling the heliconids so closely, the pierids were protected from predators, too, although they were perfectly good to eat. The birds were "fooled" into thinking that the pierids were really distasteful heliconids. Bates' discussion of mimicry was so clear and logical that it is still the starting point for the study of mimicry, even though we know much more about this complex and fascinating phenomenon today.

In 1878, a very important addition to the study of mimicry was made by a German scientist named Fritz Müller. He had also studied Amazonian butterflies. Bates had noticed that often two bad-tasting butterflies resembled one another, but could not explain why. Müller provided the answer by reasoning that predators must learn which butterflies taste good and which do not. When enemies sample them, some butterflies will be killed or injured. If there are two different distasteful species with similar patterns, each species will lose fewer individuals to the beaks of unwary birds before the birds learn to avoid them. In this way, distasteful species gain additional protection by sharing a warning pattern. Once predators learn the pattern by unpleasant experience, all similar creatures are left alone.

Mimicry Evolves

When we see the amazingly close resemblances between mimics and their models, it seems impossible to understand how they came to look so much alike. How could white or yellow pierids with wide wings give rise to black, yellow, and red butterflies with long wings? It is hard to imagine what the intermediate stages were like. The changes in species which occur over time are called evolution, and some knowledge of evolution can help us to understand how mimicry comes about.

Evolutionary changes usually occur slowly, over many generations. With time, the different species of living things change, and new species come into being. (A species is a group of very closely related individual animals or plants able to breed with one another to produce healthy offspring.) As the world around them varies, species of animals and plants become altered to adapt them to different environments. Differences in climate and in the other living things around them can drastically influence the evolution of species. The changes do not happen "on purpose"; they come about by the slow accumulation of random changes in the hereditary traits. If one of these chance changes helps an animal survive long enough to reproduce, the change is passed on to the next generation. Over many generations, species can come to look very different from their ancestors.

The evolution of mimicry has been a lively topic with scientists ever since the phenomenon was first discussed by Bates. For a long time there were scientists who did not believe in mimicry at all. They felt that some environmental factor such as climate produced the similarities between mimics and models. They did not think that birds and other predators attacked butterflies, so they did not believe that "copying" a supposedly distasteful butterfly had any meaning. But we now know from many scientific studies that butterflies are eaten by birds and other animals such as toads, and that such predators can be fooled by the similarities between mimic and model. Experiments also show that even a slight resemblance between a good-tasting insect and a bad-tasting one can be of some advantage to the tasty one. If birds learn that black butterflies with red markings should be avoided, some of the birds will also ignore plain black butterflies, even if their wings are different in shape. So an insect with a slight resemblance to a distasteful species will have a better chance of surviving than one which looks different. It will also have a better chance of reproducing and passing on the inherited traits which give it the resemblance.

But some of the predators can be surprisingly discerning, so very exact resemblance is the safest kind of insurance for the insect. When scientists feed butterflies to birds, some birds can distinguish even very close mimics. These sharp-eyed birds will eat the mimics and leave the models alone. Any descendants of a crude mimic which carry traits improving the mimicry will have increased chances of surviving and reproducing. With each generation, any new features which improve the "copying" will tend to be passed on. Thus, over time, the mimic species will come to look more and more like their models. In tropical regions, where there are many predators on butterflies, lots of them sharp-eyed birds, the similarity between mimic and model can become so exact that a scientist often cannot tell which butterfly is which until he has caught them and examined them very closely.

Even models and mimics from very different insect groups may be so similar that they fool the experts. Tiger beetles have a very nasty bite which can injure small predators. In Borneo, one kind of tiger beetle is copied so exactly by a grasshopper that for years they were both placed together in the museum collection as beetles. One day a scientist examining the collection discovered the grasshoppers by chance. Only then, by carefully sorting the specimens, could the beetles and grasshoppers be separated from one another.

Kinds of Mimicry

So far we have talked about only mimicry of distasteful or dangerous animals. While some scientists like to limit the word "mimicry" to such cases, this leaves out too many fascinating creatures, such as the cuckoo bird. Cuckoos lay their eggs in the nests of other birds and leave them there. The eggs have the same color and pattern as those of the host bird, who mistakes them for her own and "adopts" them.

Some predators fool their prey by using mimicry. Several kinds of fishes lure unsuspecting victims to their doom by displaying part of a fin which looks for all the world like a tasty morsel of food. The poor fish that thinks it will get a meal becomes a meal instead. Other predatory fish join schools of harmless fish and thus become inconspicuous. From the safe haven of the school, they dart out and bite other unsuspecting fish as they swim by.

Even plants may be mimics. Some weeds are impossible for farmers to eliminate because they look so much like the crops among which they live. A weed called gold-of-pleasure grows in fields of cultivated flax. While its wild relatives are small and branchy, the gold-of-pleasure plant grows tall and thin like flax. Its seeds are similar to flax seeds in size and weight and cannot be separated easily from them. So whenever farmers harvest flax seeds, they also harvest gold-of-pleasure seeds. When they plant those seeds, the gold-of-pleasure will come up right along with the flax, carefully tended by farmers. Other plants show a completely different sort of mimicry. Many orchids have flowers which copy female bees. Males are "fooled" into trying to mate with them. In the process, the orchid flowers are pollinated.

Another Kind of Disguise

Camouflage is different from mimicry. A mimic looks so much like its model that another living thing confuses the two. But a camouflaged animal looks as if it is not there at all. Some moths look like dead leaves and blend into the litter on the forest floor. Others have patterns which melt into the background on tree trunks. Stick insects look like dead twigs and thus may go unnoticed. Most scientists consider camouflage to be separate from mimicry. But there are some cases which make it hard to draw the line between the two. A crab spider in Malaysia is camouflaged as a bird dropping. Its body looks like the black-streaked whitish center part of the dropping, while its web resembles the more liquid spread-out part which has dried. At first thought, one might think that this is strictly a case of camouflage; the spider is not eaten by birds because it looks like a bird dropping. But the spider preys upon butterflies and other insects which are attracted to bird droppings because of the water and vital salts which they contain. Thus this is clearly a case of "aggressive mimicry," in which a predator is fooling its prey and attracting it.

Colors That Say "Stop!"

Although scientists are aware of these other types of mimicry, they have concentrated their research efforts on

A mourning sphinx moth against tree bark is a good example of camouflage.   DR. PHILIP CALLAHAN

the more common sorts. Because it was discovered by Bates, mimicry in which the mimic is harmless and the model is distasteful or dangerous is called "Batesian mimicry." When two or more kinds of dangerous or distasteful species resemble one another it is called "Müllerian mimicry," after Müller. Both of these kinds of mimicry depend on the color vision and memory of predators.

Certain color combinations seem to be easier for predators to remember and associate with unpleasant experiences. These colors are used by warningly colored animals the world over and are called "warning colors." Yellow and black or red and black are common warning color combinations. Yellow and black stripes are probably the most common warning pattern in the animal world, since bees, wasps and hornets share them. The poisonous caterpillars of the cinnabar moth display striking black and yellow rings, too. It is debatable whether these caterpillars should be considered Müllerian mimics along with bees and wasps. It is not possible without experiments to determine whether both wasps and cinnabar caterpillars have simply hit upon an easily remembered pattern or whether predators generalize the pattern from one to the other.

Batesian Versus Müllerian Mimicry

Batesian and Müllerian mimicry have very different evolutionary effects. When two or more distasteful or dangerous Müllerian mimics resemble one another, it is an advantage for them to look as much alike as possible. The closer their resemblance to one another, the easier it is for predators to remember their common pattern and avoid them all. Batesian mimicry is another case entirely. The model is distasteful and advertises the fact by its warning colors. But the mimic is perfectly harmless. If an inexperienced predator first encounters the harmless mimic, it thinks that such insects taste good; next it may attack and injure or kill one of the models. Evolution of the mimic will always be toward a close resemblance to the model. The more it resembles its unpleasant model, the more predators it fools. But the more Batesian mimics there are, the greater are the chances that predators will kill more models. For this reason, the model will tend to evolve *away* from resemblance to its mimics. A poisonous animal has better chances of survival if it lacks close, harmless mimics.

This results in an "evolutionary race," with the model perhaps changing in pattern and the mimic constantly evolving toward the changing pattern of its model. In Müllerian mimicry there may be just about any number and proportion of the species involved, since all are nasty. But in Batesian mimicry there must always be a certain minimum number of models to "educate" predators into avoiding the common pattern. Bates thought that there had to be many more models than mimics for mimicry to be effective, and in fact one does usually find an excess of models over mimics in nature. However, recent experiments led to the surprising conclusion that if the model is distasteful enough, there can be more mimics than models. If a blue jay gets very sick after eating a monarch butterfly, it will be careful for weeks to avoid any butterfly which reminds it of its awful experience.

All cases of mimicry involving distasteful insects are not

easy to place firmly in one category or another. The actual palatability of most insects is not known. Finding out just which insects taste good to birds, toads, and other animals is not easy; inexperienced animals must be tested in the laboratory, or incredibly long hours must be spent in the field waiting for the chance observation of predators testing their prey. Therefore, much of the discussion of edible versus distasteful insects is based on assumption rather than on careful experimentation.

Distastefulness can vary greatly, too. While one Müllerian mimic might taste simply awful or be downright poisonous, another may have a mildly unpleasant flavor. And, mixed in among the varyingly distasteful butterflies may be a perfectly luscious species which is a Batesian mimic of the Müllerian mimics around it.

Odor, Touch, and Sound Mimicry

Mimicry need not be visual. While we use mainly our eyes to get an idea of the world around us, other senses are more important to some other animals. Ladybird beetles taste bad and have a characteristic odor which is "copied" by some tiger moths. Termites live in dark nests and rely on smell and touch to perceive their environment and to recognize one another. Many beetles which take advantage of the protection offered by a termite nest look nothing like termites, but probably smell and feel enough like them to be accepted. Some beetles which live with ants seem to mimic only key features of ants. One kind, for example, has a small part of its body which has the same texture as the body of ant larvas. If a worker ant touches this part of the beetle with its antennas, it picks the beetle right up and carries it off just as if it were an ant larva. When the harmless eastern indigo snake is disturbed, it hisses and vibrates its tail, producing a rattlesnake-like rattling sound. This behavior presumably scares off predators.

Unfortunately, very little scientific study has been done on these other types of mimicry, partly because of the limits of our human senses and partly because of the difficulty in working with animals such as ants, termites, and snakes in the laboratory.

2 · Saga of the Monarch and the Viceroy

People have always been fascinated by the grace and beauty of butterflies, flitting through the air so freely. Scientists are not immune to the charm of butterflies, and so these animals have been collected and studied more than many other creatures. Mimicry was first discovered among butterflies, and the familiar monarch and viceroy butterflies have provided scientists with a convenient example of Batesian mimicry for experiments.

The majestic monarch is one of the most familiar American animals, and the smaller but similar-looking viceroy is also known to many people. But how many have thought about why these two insects should so resemble one another? Scientists have debated about the monarch and viceroy for years, some feeling that the monarch was distasteful to birds and that the viceroy was a mimic of the monarch, while others did not believe that mimicry was involved at all. As recently as 1957, one scientist proposed that birds avoided eating both butterflies because both had characteristics in common which somehow kept birds from trying to eat them. He suggested that birds might see the flying butterflies as dry leaves being tossed about by the wind rather than as possible food. It was not until the late 1950s that laboratory scientists finally began to study mimicry. The distastefulness of the monarch and its probable mimicry by the viceroy were the first problems dealt with.

Are Monarchs Distasteful?

The larvas of the monarch butterfly feed on various plants in the milkweed family. Many of these plants contain potent poisons. The larvas which feed on these plants store up the poisons, so the adult butterflies also contain them. No only do the poisons make the monarchs taste bitter, they can also make birds which eat them throw up, an unpleasant experience which a bird remembers and associates with the food which caused it. Jane Brower was the first scientist to show that inexperienced caged blue jays would eat viceroys but, after trying monarchs, would avoid both butterflies. Other scientists, however, fed monarchs to various birds and found that some monarchs were eaten by the birds without bad effects.

This confusion over whether or not monarchs are poisonous has been cleared up. Monarchs which feed as larvas on poisonous milkweed are poisonous as adults, whereas those which feed on nonpoisonous milkweeds are perfectly edible as adults. Jane and Lincoln Brower devised a way of measuring the potency of monarchs. They ground up the butterflies and force-fed capsules contain-

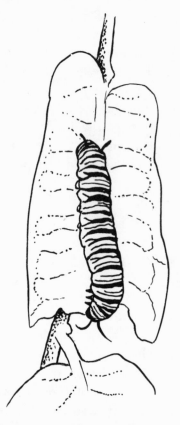

A monarch larva, its head at the bottom of the picture, eats away a milkweed leaf. The poisons in the leaf do not affect the caterpillar, but for birds that eat either the larva or the adult it's a different story.

ing measured quantities of monarchs to blue jays. They then determined how much of the butterflies it took to make a blue jay throw up. They found that monarchs reared on some milkweeds were six times as potent as those reared on others. Field collections of monarchs gave interesting results as well. Only 24 per cent of the monarchs collected in Massachusetts made birds vomit, while 65 per cent of Trinidad monarchs did. No wonder some scientists found that birds would eat monarchs, while others found they would not. It depended on which population of monarchs was used in the experiments.

Beginning with what seems like a "simple" case of mimicry, we have already run into complications. Yes, monarchs are poisonous and warningly colored, and viceroys are mistaken by birds for monarchs and are avoided by experienced birds. But monarchs which are perfectly harmless also mimic the poisonous monarchs. The poisonous butterflies protect not only viceroys but edible monarchs, too.

Powerful Poisons

The poisons which monarchs obtain from milkweed, called cardiac glycosides, are powerful indeed. Like many poisons, they have medical uses when given in small, controlled doses. In humans, they strongly affect the heartbeat and are used to treat people with heart disease. A

Blue jays tested in the laboratory, like this one, had vomiting spells after eating monarch butterflies and thereafter refused to touch one. In the wild, viceroys thus benefit from their resemblance to monarchs.

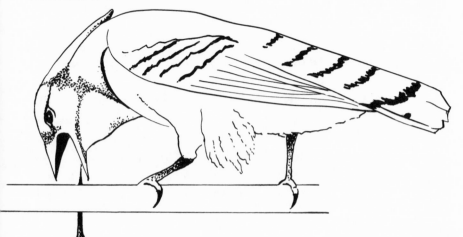

monarch raised on a potent milkweed contains more than the amount of similar glycosides given to someone on the verge of dying from a heart attack. As a matter of fact, such a harmless-looking butterfly has almost enough poison to kill a person. It is fortunate that no tragedies resulted from the casual tasting of mimetic butterflies by early naturalists intent on finding out which butterflies were "distasteful."

A poisonous monarch contains more than one kind of cardiac glycoside, and these chemicals vary in their potency. Small quantities of some will make a bird very sick, while much larger amounts of others have no apparent effect. Their degree of bitterness also varies. The Browers, working with other scientists, have taken monarchs apart to see if there are variations in the distastefulness and potency of different butterfly body parts. They found that the concentration of glycosides was highest in the wings and lowest in the thorax, while the abdomen had an intermediate concentration. But they also found that the potencies of the glycosides from different parts of the body varied. Those in the abdomen were the most potent; a blue jay which ate an abdomen from a toxic monarch would get very sick. The wings from the same butterfly would make the bird less sick, even though they contained more glycosides.

These findings are interesting when combined with knowledge about how a bird eats a butterfly. If an inexperienced blue jay is given a nontoxic monarch, it carries the butterfly to its perch, pecks off the wings and legs, and then swallows the butterfly's body. The first poisonous butterfly goes the same way; apparently the bitterness is

not strong enough to make the bird reject the butterfly
when it has no choice of food.

But after the bird throws up, things change. If it attacks
the next monarch at all, it will take the butterfly to its
perch and move it around cautiously with its bill and feet.
A jay which is fed nothing but monarchs can learn right
away to tell poisonous monarchs from nonpoisonous ones,
probably from tasting the high concentration of bitter
glycosides in the wings. It will eat the nonpoisonous ones
and reject the toxic ones. However, if the same bird has
a choice between any monarch and a harmless butterfly
which looks different, it will reject the monarch on sight.

Wild birds have a choice of food, but even so they may
occasionally "sample" a monarch to see if it is bitter. When
collections of butterflies are made, more specimens of un-
palatable kinds show beak marks on their wings than do
individuals of edible kinds. This could happen when birds
are especially hungry and desperate to find food. Or it
could be a consequence of Batesian mimicry; since not all
butterflies which look like toxic ones actually are toxic, it
may be worthwhile for a predator to sample warningly
colored individuals now and then to see if they are truly
distasteful.

Some insects which are to be avoided, either because of
their taste or their painful sting or bite, may be able to
survive an attack by an inexperienced predator. Because
their warning colors make them conspicuous, they are
more likely to be attacked at first than are camouflaged
insects. Warningly colored insects often have very tough
wings and bodies which improve their chances of surviving a "taste test" by an unwary predator. Scientists have

found that birds often peck at their prey ten or more times before killing it, giving the bird plenty of time to discover that this insect is to be avoided rather than eaten. In the wild, this may be what happens to monarchs. While a blue jay in the laboratory will eat a bitter monarch the first time around when it is the only food provided, a wild bird might take one mouthful of bitter monarch wing and decide on the spot to reject it in favor of more pleasant eating.

The Cost of Carrying Poisons

By now we may wonder why more insects are not protected from enemies by poisons. If being distasteful works so well for the monarch and other butterflies, why shouldn't it work well for all? A poisonous animal has one obvious problem to deal with—how to keep from poisoning itself with its own defense. This is not an easy matter; it takes energy to guard the body of a monarch from its own poisons, and the more poisonous monarchs apparently do not live as long as the less poisonous ones.

Like many other toxic butterflies, monarchs are long-lived. But they inhabit North America where winters are often severe, and they have no way of protecting themselves from the cold. Monarchs avoid winter by flying south, to the mild California coast or to Mexico. This migratory behavior is completely instinctive, for the butterflies which grew up in the North follow traditional flight paths to established wintering spots in the South which they have never seen. When monarchs are captured at various points along the migration route, the percentage of toxic butterflies decreases as the monarchs fly south. Many of the highly toxic individuals appear to die along the way, while more of the less poisonous ones make it to the wintering grounds. And toxic monarchs tend to be smaller and weigh less than edible ones. Thus it seems that being poisonous has its disadvantages.

Female monarchs overall are 24 per cent more potent at making blue jays throw up than males are. Why should this be? In order to reproduce, a male butterfly must only mate before dying. But a female butterfly must live long enough to lay her eggs. And when she lays them, she is an easy target for hungry birds. The female monarch flies slowly about open fields, searching for milkweed plants. She must land on the plants to lay her eggs, and could so easily be picked off while doing so. Despite the difficulties of dealing with the poisons in her body, the female monarch gains enough extra protection from being highly toxic to make it worthwhile.

California monarch populations have many more completely palatable butterflies than Massachusetts populations. In California, almost half of them are quite edible, while most of the rest are especially toxic. Only 10 per cent of the Massachusetts monarchs are completely harmless, but none of the toxic ones are as potent as the poisonous ones in California. A few highly toxic butterflies can be effective in deterring predators. If a bird gets very sick after eating a butterfly, it will give a clear berth to any similar food for a long time. But if it becomes only mildly ill, it might be willing to sample such butterflies again if it is hungry. Because of their lower toxicity, more Massachusetts butterflies must carry at least some bitter gly-

cosides. But in California, the few very poisonous butterflies can protect the many edible ones.

Enter the Queen

There are still further complications in the monarch-
viceroy story, for they are not the only large, orange-
winged butterflies in North America. There is also the
queen, a relative of the monarch. Basically a tropical species, the queen is found in the Caribbean, Florida, and
Texas, as well as in Central and South America. Queens
from different areas look quite different, but they are all
warningly colored with orange or dark red and black.
They contain poisons like those of the monarch. In Trinidad, both queens and monarchs are found, and the queens
there look very much like monarchs. Since only about 15
per cent of Trinidad queens are poisonous, the many nonpoisonous queens are mimicking both monarchs and poisonous members of their own species.

Queens in Florida look different from those in Trinidad
and apparently are more poisonous. Monarchs are quite
rare in parts of Florida and southwestern states where
queens are found, but viceroys do live there. Since viceroys are perfectly edible, they would be easy targets for
predators in Florida if they looked like monarchs. The
predators would not have encountered distasteful monarchs to discourage them from eating viceroys, so they
could feast on the brightly colored butterflies with no
hesitation. But viceroys in Florida are still protected, for
they look like queens instead of monarchs. The Florida
queens are a mahogany color instead of orange, and so are

the Florida viceroys. Over time, Florida viceroys which looked like monarchs got eaten more often than those with markings and coloration more similar to queens. The more queenlike viceroys left more offspring with each generation, until a different race of viceroys was produced. As we will see, the same sort of thing has happened with many butterfly species; some species have a dozen or more different forms, each mimicking a different toxic butterfly.

3 · Beauty on the Wing: Tropical Butterflies

Butterflies abound in the tropics. Tropical butterflies come in many sizes, some very small and delicate and others with heavy bodies and wide wings. The colors of tropical butterflies provide a breathtaking array of beauty, with iridescent blues, brilliant reds, and vibrant yellows common. While many tropical butterflies are disguised by their colors, others fly casually by, displaying their bright wings with no fear. These are the distasteful butterflies and their mimics—dozens of kinds of models and mimics from many different butterfly families.

All members of the butterfly family Danaidae appear to be distasteful. The monarch belongs to this family, but most of its relatives live in tropical places. Because of their flavor, danaids often have mimics. One African danaid which has mimics is called the friar. This rather pretty butterfly is black with large patches of white on all its wings. There are also small white blotches near the forewing tips. One mimic of the friar is a type of admiral butterfly. The admiral looks almost exactly like the friar. The shape of the wings is almost identical, and the color pattern is essentially the same. But, as is usually the case in such Batesian mimicry, the friar and its mimicking admiral belong to completely different butterfly families. While the friar is a distasteful danaid, the admiral is a perfectly edible brush-footed butterfly which can be easily identified by its short front legs.

More African Mimics

This admiral is not the only butterfly to mimic the friar and its relatives. Probably the most amazing mimic species in the whole animal kingdom, the African mocker swallowtail, has forms mimicking several distasteful danaids, including the friar. Thus a scientist collecting butterflies on the African east coast may find butterflies with almost identical wing patterns which belong to three quite unrelated butterfly families.

Swallowtails usually have extensions on the hind wings which look like tails. All African mocker males are typical swallowtails. Their wings are mostly creamy white with some black markings. The females, however, come in a bewildering variety of forms. Some resemble the friar and its kin, while others copy the African queen butterfly (a close relative of the American queen and American monarch). Still more look like other distasteful butterflies from other families, while some have the same pattern and appearance as males of their own kind. There are also forms of the mocker which are unique and do not mimic any other butterfly.

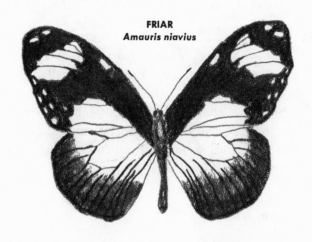

FRIAR
Amauris niavius

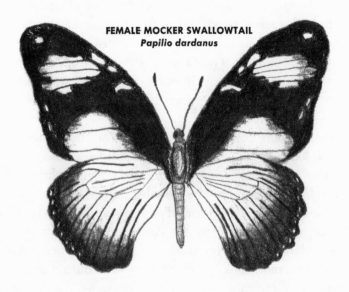

FEMALE MOCKER SWALLOWTAIL
Papilio dardanus

The admiral at upper right mimics the friar butterfly, though it is of quite a different family. The female mocker swallowtail mimics the friar, and also butterflies of various other

ADMIRAL
Hypolimnas dubius

MALE MOCKER SWALLOWTAIL
Papilio dardanus

families. Some of these mocker females look like their own males, which have the "swallowtails" on their wings. Drawing by the author.

Why are there so many forms of the mocker? And why should only the females be mimics? Wouldn't it be helpful to the species if the males could be "disguised" as well? While scientists have discussed these problems for years, they have yet to agree upon answers which satisfy them all.

Bates originally thought that models had to outnumber mimics for his kind of mimicry to work. Scientists have since found that very unpleasant models can protect quite a large number of mimics, but there are limits. The more models there are, the greater the chance a predator will encounter a distasteful butterfly before a tasty one. This limits the population of mimic species. There can never be too many mimics of any particular model, or the mimics will not be protected. Species like the mocker seem to have "solved" this problem of limited numbers by having many different forms mimicking different models. Several forms of this swallowtail may live in one area, each form with a different model. There can be large numbers of mockers present, but all are protected.

Some scientists believe that all mocker males must be alike so that the females can recognize them on sight. If the males came in a variety of forms, the females could not distinguish males of their own kind. But no one has studied the mating behavior of the species, so we do not know how the males and females recognize one another. No butterfly species yet studied has females which are attracted to males on sight. Male butterflies keep a lookout for females, and tend to go after anything remotely resembling a butterfly. Only when they get close do they decide whether to court or not. And with many butterflies, odor is much more important than sight in mating, so this explanation is probably unsatisfactory.

As we saw with the monarch, female butterflies are more exposed to predators than are males. After a male has mated, he can be eaten and his offspring will still live. But the female, slow and heavy with the load of eggs she carries, must expose herself to predators as she lays her eggs. Even a camouflaged female is in danger then, during the most crucial part of her life. But a mimetic female would be avoided by predators, even though she is so vulnerable. Because Batesian mimics must limit their numbers and because females are more in danger than males, many scientists believe that species like the mocker are able to exist in larger numbers because their females are mimetic and their males are not. There are no Batesian mimics with mimetic males and nonmimetic females, a fact which lends further support to this theory.

Mimicry Rings

When more than two species are involved in a mimicry situation, it is called a "mimicry ring." A mimicry ring may be purely Batesian, with one distasteful model and two or more edible mimics, or it may be totally Müllerian, with more than two distasteful butterflies looking alike. Many mimicry rings are extremely complex, with three or more Müllerian mimics being copied by several Batesian ones. Such collections of similar species are found most frequently in tropical regions, where the variety of butterfly life is greatest. Warningly colored butterflies often join together in large flocks, often at mud puddles or sleeping places. In some areas, such a gathering of butterflies which at first sight seem to belong to one species may contain half a dozen species from four or more families. Moths may even be present, "disguised" as butterflies. Although many individuals gathered there may be distasteful, others could be perfectly edible. Given the confusion of wings and bright patterns, it seems understandable that a predator would hesitate to single out any individual to attack.

While one mimicry ring may involve butterflies from four or more families, some species may be from the same family. This presents a problem to biologists, since related species occurring together are supposed to evolve to be as different from one another as possible. Related species must remain separate and not interbreed. If they do mate, they will either produce sterile offspring or become blended into one species. So how can closely related butterflies in the same area mimic one another and still remain separate? Studies on a remarkable pair of butterfly species from South America help provide answers but also raise even more questions and provide more puzzles to the student of mimicry.

The Mysteries of Passionflower Butterflies

Passionflower butterflies live mainly in the tropical forests of Central and South America. Like so many other butterflies, their unpleasant taste protects them from becoming bird food, and their narrow black wings are marked with spots and bands of yellow, white, and red, like other warningly colored animals. They fly with a slow, floating motion which shows off their bright colors. They often gather together to rest, and they are mimicked by butterflies from at least three other families. So far, passionflower butterflies appear to be just another example of the kinds of butterflies found in Africa which we have already learned about. But there are important differences. While only 40 or 50 species of these butterflies exist, about 750 scientific names have been given to the incredible variety of forms which have been described. In some areas, one species may produce 50 different varieties, each with its own distinguishing name. But while one species may have dozens of forms, two different species found in the same area may have identical forms. How can we make sense of this paradox?

On the island of Trinidad lives a black butterfly whose only marking is a bright red patch on each forewing. It would seem that all the butterflies with this pattern would belong to one species; all are passionflower butterflies. But the identical adults are produced from two different kinds of caterpillars which feed on different food plants. Clearly, two species of butterfly are involved which are Müllerian mimics of one another. The bright red patches on their forewings warn birds which have tried either species to avoid both in the future. These two kinds of passionflower butterflies are named *Heliconius erato* and *Heliconius melpomene*. Both species are found over vast regions of South and Central America and exist together in many areas.

The amazing thing about *H. erato* and *H. melpomene* (scientists often abbreviate the first, or genus, name when using scientific names) is that wherever they coexist, they look so much alike that even scientists may have trouble telling them apart. But they do not always have the simple

Two species of the passionflower butterfly: Heliconius melpomene *at top and* Heliconius erato *below. The two insects shown are from races that have minor differences, but in many other cases the two species can hardly be told apart even by experts.*

color pattern found in Trinidad. In fact, over their common range, they have about a dozen shared patterns which may vary greatly from the Trinidad one. In Ecuador both butterflies have forms with two large white spots on their forewings and only a hint of orange, while in Bolivia both have a pattern with yellow spots on each front wing, large orange patches on all four wings, and rays of orange on the hind wings.

It is easy to see why these two butterflies would come to resemble one another in one area. By sharing the same warning pattern, the two species also share the burden of being tasted by inexperienced birds. The birds need learn only one warning pattern instead of two. But why do different forms of the two species exist in different areas? Why isn't there one common pattern shared by both species wherever they are found?

A British scientist named John R. G. Turner has studied these butterflies for years. He believes that the answer to this question lies in the geological history of South America. At present, most of South America is covered by rain forests. But during the last Ice Age, periods of drought occurred which isolated areas of forest into "islands" surrounded by drier regions with few trees. Since Heliconius butterflies need the forest, they, too, were isolated into separate populations which could not breed with one another. The butterflies in each little forest "island" evolved on their own, independently of those in other isolated forest areas.

We can imagine that *H. erato* and *H. melpomene* already existed before the last Ice Age and successfully inhabited the forest together. They probably already resembled one another, and their color pattern may have been similar throughout their whole range. Since both a red splotch on the forewing and a yellow bar on the hind wing are found in several forms of the two species, perhaps this was the original common pattern. But when they became isolated into the different little islands of forest, each population was subjected to a somewhat different collection of other butterfly species. Let us say that in one of these isolated areas there was another species of distasteful butterfly with a similar pattern, but this other butterfly was more abundant than the other two species combined.

In a situation like this, predators would be most likely to find out that the color pattern of the most common distasteful butterfly was to be avoided. Any individuals of other species which might be mistaken for the most abundant kind would have a better chance of remaining uneaten. They would be more likely to reproduce and pass this similarity on to their offspring. In this way, over the 1500 to 7000 years of the last Ice Age, within each forest island, all related distasteful butterflies may have come to exhibit the pattern originally possessed by the most common species. In different areas, different color patterns would become the common one, and different races of butterflies such as *H. erato* and *H. melpomene* would be formed.

After the Ice Age ended and the forests again spread over the land, these different forms of *H. erato* and *H. melpomene* again came into contact. The forms of *H. erato* had evolved different color patterns, but they could still breed with one another. The same was true for *H. mel-*

DR. JOHN R. G. TURNER

This is also Heliconius melpomene, *of a race from the lower Amazon Basin; the wing shape and coloring are very similar to those of other of its species but marks differ considerably. Through the endless variations introduced through evolution, an insect can in time develop an appearance closer and closer to that of another, until it resembles it almost exactly.*

pomene. When two forms with different color patterns bred together, the genetic traits which each had accumulated during its long isolation combined upon cross-breeding to result in an explosion of new patterns. This would explain why there is such an incredible number of different color forms of both *H. melpomene* and *H. erato* where two mimic forms overlap.

If *H. erato* and *H. melpomene* always look so much alike, how do they tell each other apart? When it comes to mating, males and females of each species must be able to recognize one another. While these butterflies look alike, they smell very different. *H. erato* smells like witch hazel, while *H. melpomene* reminds humans of fried rice. Since odor is very important in the mating behavior of many butterflies, these two species probably have no trouble at all finding appropriate mates.

One More Puzzle

Most experts believe that warning coloration must be dramatic and conspicuous. Some feel it is important for distasteful animals to look as unlike their edible cousins as possible. Since camouflage is usually green, gray, or brown, warning colors are black, yellow, and red. But this may not be the whole answer, for it does not explain one of the important puzzles of tropical butterfly mimicry.

We have seen how distasteful butterflies in a certain area will come to resemble one another and become Müllerian mimics. Why, then, are certain kinds of warning colors associated with certain forest habitats? A scientist named Christine Papageorgis investigated the mimicry rings present in three geographically distinct areas of Peru. In each of these areas, five quite distinct mimicry rings were present, and each area had the same five. One ring consisted of butterflies with orange and black tiger-stripe patterns, while another was composed of transparent, bluish butterflies, and so on. Butterflies in each complex tended to live in a particular part of the forest. For example, the transparent ones flew close to the forest floor, while the tigers flew above them.

Dr. Papageorgis thinks that the color patterns in each mimicry ring blend in with the lighting at different levels of the forest. She believes that the mimic patterns do not attract attention, but rather help to camouflage the butterflies in flight. This would protect them from pursuit by a naive or forgetful predator. Effective warning patterns need not make the butterflies obvious and attract attention, according to this idea. They need only be easy to remember.

While these theories are interesting and may well be correct, it is also easy to argue against them. Since almost all transparents belong to the same butterfly family, they may have similar coloration because of their relatedness. The same can be said for the "orange" and "blue" complexes. Only the tiger complex appears to have members from more than two families. And because butterflies from these complexes fly at different levels of the forest, they may rarely encounter one another. They could be just as isolated from each other as species found miles apart. This could also account for their pattern differences. In order to understand why different mimicry rings exist in the same areas, other forests in other countries must be studied to see if the same kinds of coloration are always correlated with the same kind of forest environment. Certainly the idea that warning colors need not make an individual conspicuous is an idea worth considering.

4 · How To "Fool" Your Food

Hunters as well as the hunted can use mimicry to their advantage. Some predators disguise themselves as something attractive to potential prey and lure unsuspecting victims. This is called aggressive mimicry. It is sometimes difficult to draw the line between aggressive mimicry and camouflage.

We would say that a green praying mantis sitting absolutely still on a stem, blending into its surroundings and waiting for careless insects to venture close, is camouflaged. But what about the praying mantis' close relatives such as the Malaysian flower mantis, which "disguises" itself as an attractive flower? These insects can become pink or white, depending on the color of the flowers they associate with. The upper parts of their second and third pairs of legs, as well as the thorax and abdomen, are flattened and resemble petals. The mantids sit among the flowers and grab bees and other insects which come to gather pollen and nectar from the flowers. We could easily call this camouflage, for isn't the mantis hiding among the flowers and blending with them? But on the other hand, the mantis may be attractive to insects even when it sits alone among the leaves. If so, it is not just hiding among the flowers, it is "mimicking" a flower and doing a good enough job of it that its prey is fooled and actually attracted.

Another kind of flower mantis lives in Africa. The young mantises become associated with plants with red or greenish flowers. As they grow up, they develop colors which match their chosen background. They blend in so well that they are hard to see even if you know they are there. If red mantises are placed on greenish white flowers, or vice versa, they can gradually change color to match their surroundings, but for several weeks they are a conspicuous pinkish green color.

Fishing Fish

Often in aggressive mimicry, only a part of the hunter is modified to "fool" its prey. The anglerfish are probably the best-known lurers. There are many different kinds of anglers, and each has a lure which is suitable for attracting particular kinds of prey fish. The lure is an enlargement of a spine on the top fin. Many anglers have a worm-shaped bit of tissue attached to the fin by a thin stalk. When a hungry worm-eating fish comes along, the angler bounces and jiggles the lure, attracting the hopeful prey within reach of the angler's huge mouth. When a fish is enticed close, the angler opens its mouth very suddenly, creating a rush of water into the empty cavity. The helpless prey is swept in by the current and swallowed immediately.

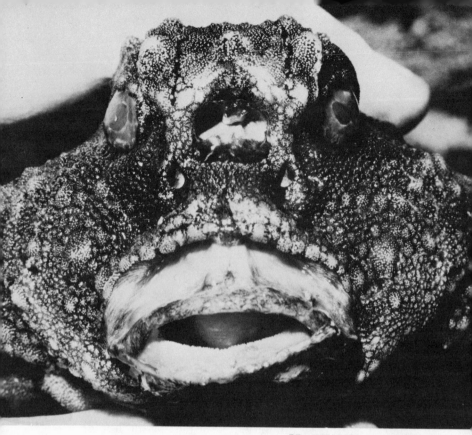

This batfish, seen head on, has a wormlike lure that it keeps in a recess until it is ready to take prey; then it dangles the lure in front of its head.

The appearance of the lure varies, depending on the habitat of the angler species. While wormlike lures are the most "popular," others resemble bundles of algae. Deep-sea anglers are among the most bizarre of them all. Their mammoth mouths are capable of taking in fish as big as themselves. Only the females have lures, for the males are very small and spend their lives attached to the bodies of the females. The lures of deep-sea anglers are suspended on especially long "lines" in front of their faces.

They live at great depths where no light penetrates, and their lures glow in the dark. Since different kinds of deep-sea anglers have lures of different shapes which glow with different colors, each is probably mimicking a different sort of food object. Unfortunately, deep-sea life is extremely difficult to observe in its natural state, and very little is known about the lives of these creatures.

Recently a different sort of luring by fish was discovered in a small Hawaiian scorpionfish. A few of these were caught and observed in the laboratory. Like other scorpionfish, these predators blend in perfectly with their surroundings. Their colors and skin texture make them difficult to see when they are lying motionless on the ocean bottom. When prey fish were added to an aquarium containing a scorpionfish, it turned its body ever so slowly until it faced the prey. Its top fins were held down, blending in with the rest of the body. When the scorpionfish faced its prey, it raised the top fin. Suddenly there was a tiny fish at that spot, or so it seemed. A black spot on the fin became enlarged and surrounded by white, resembling a fish eye. A gap between the first and second fin spines looked like a little mouth, and the long fourth spine was the lure-fish's top fin. The front of the lure-fish was a deep red color, while the back was an eye-catching yellow. A red line along the bottom of the lure separated it from the clear part of the scorpionfish's fin below. The scorpionfish snapped its fin back and forth so that it resembled a sickly small fish, easy prey to a hungry passerby. As the fin was snapped back and forth, the "mouth" of the lure appeared to open and close, just as the mouth of a real fish would do. When a bigger hungry fish approached the lure, ex-

ROBERT SHALLENBERGER/AHUIMANU PRODUCTIONS
This Hawaiian scorpionfish has what looks like a smaller fish on its back, and predators attracted to the little "fish"—actually a mimicking fin—are lured close enough to be snapped up by the scorpionfish.

pecting a meal, it became an easy catch for the scorpionfish, which immediately engulfed and swallowed it.

Do Snakes "Charm" Their Prey?

A lot of folklore surrounds snakes. Some people find them frightening and repulsive. But to a biologist, snakes represent one of the most successful animal groups, a group which exhibits many peculiar and unique adaptations to life on our earth. The folklore of many countries contains stories of snakes charming their prey in some way, and it is now known that, indeed, some snakes are able to lure their victims to their death. It is not big, powerful snakes which do so, but very young snakes of poisonous kinds. The young of many species have white, yellow, or pinkish tails. At least 23 species of pit vipers have conspicuously marked tail tips, and for many years accounts of young snakes luring prey by waving the tips of their tails have circulated. The great reptile expert Raymond L. Ditmars claimed to have seen it, as did many other naturalists. But because of the scattered, unsystematic way in which the examples were reported, most scientists felt that more firm proof was necessary to prove that luring by snakes actually existed.

Successful luring of prey with the tail has now been

Close-up of the fake "fish" shows how the first and second fin spines mimic a mouth.

ROBERT SHALLENBERGER/AHUIMANU PRODUCTIONS

observed for sure in two kinds of moccasins, and pictures have been taken. Young of the Mexican moccasin raise their yellow tail tips high above their coiled bodies and wriggle them when small lizards are put into their cages. To human observers, the tail tip comes to resemble a wriggling insect larva, and lizards apparently think the same thing, for they will approach and try to eat the tail. Instead, the young snake strikes and captures the lizard for its own dinner. Similar luring, with prey capture, has also been seen in an Asian moccasin, and young snakes of several other species have tried it in captivity. Baby copperheads, cottonmouth moccasins, and banded rock rattlesnakes all make luring attempts. The tails of some kinds make very convincing lures. The tip may be slightly swollen and so resemble an insect. It may have a dark spot at the end which looks like a head, and other markings on the tail may enhance the effect.

Adult Saharan sand vipers use the same technique. These poisonous snakes bury themselves in the sand under desert bushes during the day, with only their snouts and eyes above the surface. If a hidden viper sees a lizard approaching, it protrudes the tip of its tail above the sand. The tail tip is banded with black and white, and as the snake slowly waves it back and forth, it wriggles like an insect larva. If a curious lizard gets too close, the viper strikes and eats it.

Fatal Fireflies

While the flashing of fireflies at dusk may seem magical and mysterious to humans, it is down-to-earth business for the fireflies themselves. Fireflies are not flies at all, but beetles, and their flashing is their way of attracting mates. While the female beetles sit on blades of grass, males fly about, giving off flash patterns which differ from one species to another. The females respond to flashes from males of their own kind with a characteristic response which also varies from one species to the next. These differences in male flash patterns and in female responses allow several different species to fly at the same time in the same meadow and still find mates of their own kind easily. But for some males, the search for a mate ends in death. Also waiting in the grass, answering the flashes of firefly males, are females of another, predatory species. A predatory female sits on a blade of grass or on the ground. When a male firefly flashes nearby, she answers with a good imitation of the proper response for his species. If the male is fooled, he flashes again, and she repeats her answering signal. The male flies ever closer and lands by her. This proves to be a fatal mistake, for the predatory female then grabs him and gobbles him up. The firefly predators are very versatile, for they can successfully imitate the response patterns of several prey species.

The Imitation "Come Hither"

Most familiar spiders make large, impressive webs to catch their prey. Because of its large size, the chances are good that enough insects will accidently fly into such a web and become stuck there, nourishment for the resident spider. Scientists have been puzzled for years, however, about the success of another prey-catching technique used by spiders. Bolas spiders, found in the Americas, are closely related to web-building spiders, but they do not bother to make webs. Instead, they hang down on a thread which suspends them from nearby plants and hold a short thread with one front leg. At the end of this thread is a sticky silk ball which the spider flings at passing insects. If the ball hits an insect it sticks, and the spider climbs down the thread, stings the insect, and feeds. This seems like a very inefficient method of catching prey; how many insects would by chance fly close enough for the spider to have a chance at capturing them?

Now scientists appear to have uncovered the secret of the bolas spiders. They apparently do not rely on chance at all, but rather lure their prey by mimicking their sex attractants. When the prey captured by bolas spiders are examined, they turn out to be all male moths of only two or three kinds. Females of these species release powerfully attractive chemicals when they are ready to mate. Male moths are very sensitive to these attractants and may come to a female from far away, following the wind bearing her scent. The bolas spider appears to produce its sex attractant-mimicking chemical only while actually hunting. Spiders in a resting position do not attract moths. But when the spider prepares its sticky ball and hangs down waiting for prey, male moths soon appear, always from downwind. The moths often approach the spider several times, completely unaware of the mortal danger they are in.

Using this unique hunting technique, bolas spiders seem to catch just about the same number of victims as do web-building spiders, without having to spend the time and effort of making a web. They do, however, have to use energy to produce the chemical attractant and to swing their sticky balls at the approaching moths, so it is difficult to say if they are actually spending less energy to obtain their prey than are web-builders. There is also the problem of how young bolas spiders feed, since they weigh but a fraction of the weight of the moths attracted by the adults. Perhaps they produce different chemicals which attract smaller prey, or maybe they hunt by a completely different method. No one as yet knows.

Wolves in Sheep's Clothing

Yet another new type of mimicry used by a predator was discovered in the late 1970s. Ants often tend colonies of aphids, feeding on honeydew, a solution of excess sugars which the aphids produce. The ants "adopt" whole colonies of aphids and defend them against enemies. One predator on aphids has found a way around the alert guarding of aphids by attending ants. This green lacewing larva was discovered when colonies of woolly alder aphids were examined very closely. The woolly aphids have a very peculiar appearance, for their bodies are covered with a dense, fluffy coating of white wax produced by special gland cells in their skins. Hidden among the masses of woolly aphids lurk lacewing larvas which have removed wool from aphids and covered their own bodies with it. The larvas are about the same size and shape as the aphids and look just like them when covered with wool. The ants which care for the aphids do not notice the lacewing larvas, which can therefore feed on the aphids in protected secrecy.

If the wool is removed from the body of a larva and the insect is put with a colony of aphids, the guarding ants quickly notice it. One of the ants grabs the intruder in its jaws and drags it off, dropping it from the branch onto the ground below. If a larva already covered with wool is placed in the colony, however, the ant guards examine it and leave it alone. If an ant does venture a test bite, it quickly lets go, its mouth full of the waxy "wool." From then on, it leaves the larva alone.

The lacewing larvas have long hooks on their backs with curved ends which help hold the strands of wax in place. If a bare larva is given a chance to recover itself before the ants find it, it moves from aphid to aphid, scooping gobs of wool off with its head and pushing the strands onto its back by arching its head backward. In less than 20 minutes the larva is again disguised and protected.

The Fishy Mussels

Predators may be aggressive mimics, and so may parasites. While in theory the possibilities for mimicking by parasites are limitless, only a few documented examples are known. One striking case is found in fresh-water mussels. The female mussel broods her developing offspring in a special pouch. There they grow into strange little larvas, called glochidia, with two minute shells. In order to develop properly, glochidia must clamp onto the gills of a fish. They gain nourishment from the fish and grow into little mussels. Then they drop off and begin their independent existence.

The lure of the female mussel Lampsilis ventricosa *resembles a small fish most remarkably. It is not a lure that helps the mussel's nutrition, however, but rather its reproduction. Drawing by the author.*

The female mussel of several species has a special way of increasing the chances that her offspring will find a fish. When her brood pouch is full of developing young, she grows enlargements of her body between the two halves of her shell. At one end is a black spot, surrounded by white, which looks like the eye of a little fish. At the other end are taillike flaps. The flaps can be moved by the mussel, giving the impression of a small weak fish trying to struggle upstream. While there are always flaps of tissue which can be partially extended, only while brooding young do they become enlarged into the "fish lure" form. Larger fish are attracted by the lure. If the shadow of a passing fish falls on a brooding female mussel, she contracts her body suddenly, releasing a mass of glochidia which are stuck together. This blob is just under an inch in size, and a fish which has approached anticipating a meal is likely to snap it up. While many of the glochidia will be swallowed, plenty of them will find their way to the fish's gills instead, where they can safely grow into mussels.

Birds "Fool" One Another

By human standards, the cuckoo bird would be considered an exceptionally lazy creature. After mating, the female cuckoo seeks out the nest of another bird. She lays one egg in that nest and goes on to another, again laying one egg. When she has finished laying, she is finished with her job as a parent. The other birds will raise her young as "foster children." The young cuckoo hatches at about the same time as the host's young and rapidly takes over the nest. Any other young birds or eggs in the nest are lifted by the naked and blind but strong infantile cuckoo upon its back and dumped from the nest. Now it is the only occupant, and its "foster parents" continue to feed it, even when it grows much bigger than they are.

Through evolution, cuckoos have become experts at "fooling" their hosts. They lay eggs which resemble host eggs very closely. Different female cuckoos lay in nests of different bird species, yet their eggs always hatch. The European cuckoo parasitizes hosts such as warblers, shrikes, and sparrows. Cuckoos which lay eggs in reed-warbler nests have pale blue eggs with brown blotches, while those which parasitize redstarts lay pure bird's-egg blue eggs. Some cuckoos lay brown eggs and others pale tan ones with spots, but all match the eggs of their own host bird.

Since cuckoos cannot be raised in captivity, we can only guess at how this exact correspondence of parasite to host comes about. Each female cuckoo was raised by hosts which accepted the egg from which she hatched. The appearance of the eggs is determined by heredity. A male cuckoo mates with many females. Since these females may all parasitize different hosts, the inheritance of egg pattern and color almost certainly is carried only by the female. Otherwise the pattern could not be so exact, since the male could have been raised by a different kind of host. Then what determines how the female chooses a host which matches her egg pattern? If the egg pattern is determined only by the female cuckoo's inherited traits, then her eggs will look just like her mother's eggs. While being brought up, the female probably learns the songs and appearance of her foster parents, and simply parasitizes birds like them when she grows up. This would insure that each female cuckoo lays eggs in nests of the appropriate hosts. The occasional "mistakes" which are made are probably due to unavailability of host nests when the cuckoo had an egg to lay.

Indigo Birds and Fire Finches

Brood parasitism is actually quite common among birds. The honey guides of Africa lay their eggs in nests of wood-

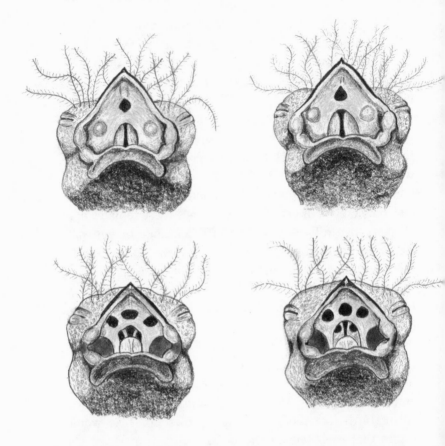

These are the heads of young birds with their mouths open awaiting food. At top left, the gape pattern (marks within the mouth) of the melba fire finch; at top right, the gape pattern of the paradise widowbird, which copies it. The inside of the mouth in each case is pinkish red, with blue circles at each side and a black mark in the middle.

At lower left is a Jameson's fire finch with its mimic, the purple combassou, another type of finch. Drawing by the author.

peckers and other birds, while black-headed ducks leave theirs for other ducks to tend. But the most remarkable of all brood parasites are the indigo birds (also called widowbirds) of Africa. Breeding males are handsome fellows, with glossy green or blue feathers. The paradise widowbird male has striking black, white, and yellow plumage, with long, graceful tail feathers. Females and nonbreeding males resemble sparrows. There are many kinds of indigo birds, and each species is a parasite of a different sort of fire finch. The adult fire finches look very different from the indigo birds. They have rather short tails and fat bodies, and most have reddish feathers. A variety of host-parasite pairs exists all the way from the edge of the Sahara desert to the tip of Cape Horn. In every case, one certain species of indigo bird parasitizes a particular kind of fire finch.

A female indigo bird probably lays only one egg in each fire finch nest, just like the female cuckoo. But the young indigo bird does not toss out its nest mates. Therefore these parasites are not as destructive to their hosts as cuckoos; the parent birds must work a little harder to fill the extra beak in the nest. Young of birds raised in nests often have bright outlines on their beaks or patterns inside their mouths. These serve as markers for the parent birds, which stuff food into their chicks' gaping beaks. But fire finch offspring go a step further. They have complicated color patterns on their beaks and mouths. When a young fire finch gapes at its parents, a pattern of spots and patches of blue, black, and red or yellow is presented. Each fire finch species has its own particular pattern, and the parents will put food only into a mouth with exactly the correct pattern. Any young bird with the wrong pattern would not be fed and would die.

To be successful, the young of the parasitic indigo birds must also exhibit the correct mouth pattern. The open mouth of an indigo bird chick has exactly the same pattern of spots and colors as those of its host. Not only that, but the indigo bird young also mimic the begging calls and begging movements of the particular host fire finch to perfection. In every important way, the chicks of the two kinds of birds appear identical to the fire finch parents. As they grow up and get their feathers, the young indigo birds match the plumage of their host's offspring as well. Otherwise, their foster parents might see them as not their kind and reject them. Only after leaving the nest do the indigo birds take on the appearance of their own kind.

When we look at such remarkable cases of mimicry, we are probably seeing the evolutionary history of a relationship between two species. Indigo birds and fire finches have probably been associated for countless generations. Although the fire finches do get to raise their own chicks along with the young indigo bird, feeding yet another mouth must put a strain on the energies of the parents. Any adult fire finches which had a way of distinguishing an alien chick in their nest would be more successful at raising their own offspring. They would add more to the next generation than birds which raised indigo bird young. The specific patterns and behavior of the young fire finches probably evolved one by one as clues to the correct identification of fire finch offspring. But the indigo birds were evolving at the same time. The indigo birds with young resembling the host most closely passed on their traits to the next generation, while those whose offspring were detected as infiltrators failed to reproduce. The resulting "evolutionary race" produced the highly specific characteristics of the young birds which we see today.

The indigo birds also mimic the calls of their fire finch hosts. When a male indigo bird sings to attract a mate, he mixes the calls of his host fire finch in with typical indigo bird calls. He learns the fire finch host's songs while growing up by listening to his foster parents. The female indigo birds become familiar with the same calls when they are young. When they are mature, they are attracted only to male indigo birds which use calls from the correct host. In this way, a female indigo bird can easily find a mate of her own kind.

5 · Mimicry in "Cold-Blooded" Animals

Fish, amphibians, and reptiles are considered to be "lower" forms of vertebrate life. They are more primitive in their body structure and appeared on earth before birds and mammals. Indeed, some extinct fish, amphibians, and reptiles were the ancestors of birds and mammals. Even though they are less advanced in some ways, these animals have been on earth longer and have evolved as many specialized and interesting adaptations as have "higher" animals.

In many ways, life along the coral reefs of tropical seas resembles life in tropical forests. Scientists and skin-diving hobbyists alike have commented on the resemblance between the brightly colored reef fish flashing through the sun-dappled sea and the equally brilliant tropical butterflies flitting through the rain forest. It should be no surprise, then, to learn that mimicry among reef fishes may be every bit as common as it is among tropical butterflies. Scientists have only begun to study possible cases of mimicry in fish, however, so most of the examples of fish mimicry are possibilities in need of investigation rather than established facts.

One fish family, the blennies, seems to be involved in many of the possible fish mimicry situations. One blenny found around the Great Barrier Reef of Australia has poison glands which give it a nasty bite. Laboratory tests of this tough customer show that unwary predators will reject the blennies alive and unharmed after trying to eat them. This fish has a long body with three dark stripes. It hovers just above the bottom and swims with short darts. At least two other fish found in the same area look and behave in much the same fashion. One is the young of a very different-looking fish, while the other is the adult of an undescribed species. One hopes scientists will soon do experiments similar to those done with butterflies to see if these possible mimics are protected from hungry predators by their similarity to the biting blenny.

Many kinds of damselfish live along the reefs. Some live protected among the stinging tentacles of sea anemones, while others swim together in schools. Among schools of one kind of damselfish swim a few young fish of another species from another family. They belong to a solitary kind, yet these young fish mingle with schools of a completely different species. Why? Since they are similar in both color and shape, they are inconspicuous when mixed in with the damselfish. Perhaps they gain protection from predators by "hiding" in the crowd. Or, since they are meat-eaters, they may be able to approach victims more closely while "disguised" as harmless damselfish.

Scientists have observed at least one other kind of fish

using schools of a different species as "cover" while approaching prey. Some blennies feed by biting chunks out of fish much larger than themselves. One of these has two different color forms. One form is orange and swims among schools of a fish called the orange anthiid. The orange anthiid is a harmless plankton-feeder, offering no threat to other fish. But the blennies which swim with them occasionally dart out and attack passing fish, catching them totally unaware. Presumably these fish are aggressive mimics, using their resemblance to the harmless anthiids as a way of sneaking up on their prey.

Cleaner-Fish Mimics

One of the most specialized life styles of reef fish is that of the cleaner fish. Each cleaner or pair has a territory, with a rock or other landmark acting as its cleaning "station." Fish of many kinds come to the cleaner. They may even wait in line for its services. The cleaner is very important to the other fish, for it removes damaging parasites and funguses from their skin.

One especially well studied cleaner is the sea swallow. This little fish has a conspicuous black line along each side of the body, and an area of bright blue near the tail. It accentuates this conspicuous coloration by performing a sort of dance, swimming forward slowly while raising and lowering its spread-out tail. Fish in need of cleaning recognize the dance and are attracted by it. While being cleaned, a fish acts as if in a sort of trance, breathing irregularly and drooping its body. It will allow the cleaner to go all over, nipping away parasites and dead skin. Many fish will spread their gill covers and open their mouths to make the cleaner's job easier. Even predators which eat other fish of cleaner size allow cleaners to enter their mouths and come out again unharmed. This is all the more remarkable since cleaners are perfectly edible and have no way of defending themselves.

Sometimes another fish is found near cleaning stations. Although it is a blenny and therefore not at all closely related to the cleaner, it looks almost exactly like it and performs the same strange, special dance. But its behavior toward a fish in need of cleaning is quite different. After the unsuspecting customer positions itself carefully for cleaning, the little mimic will bite off a big chunk of fin. The injured fish wheels around instantly, but all there is to see is the supposed cleaner, quietly staying in its place as if nothing had happened.

Fish are capable of learning, however, and the first thing they learn is to avoid the place where they had an unpleasant experience. Thus older fish will shun the phony cleaning station of a blenny mimic, while young fish and wanderers which do not stay long in one place will be fooled. With time, however, some fish can learn to distinguish the true cleaner from its mimic. While the appearance and inviting dance of the two fish are virtually identical, the behavior which follows is not. In order to bite without being seen, the mimic must approach from behind and usually ends up getting a piece of the tail fin. True cleaners, however, will approach a customer from all sides. When a fish is bitten by a cleaner mimic and turns toward it, the mimic turns away. In order to tell a genuine cleaner from a mimic, all a fish has to do is turn toward it and observe its reaction. Fish in aquariums which have had experiences with mimics and true cleaners will do exactly that and can apparently figure out whether a cleaner is genuine. Whether fish in the wild use the same method is not known, but we do know that adult fish can and do learn to tell the two little fish apart.

Because they are being subjected to such close scrutiny by their victims, we could expect cleaner mimics to resemble their models especially closely, and they do. The sea swallow cleaner is found over a wide area and has different color variations. Wherever there is a difference in the color of the cleaner, the mimic shows the same variation. For example, cleaners in one island area have orange-red spots on their sides, and the mimics have spots of the same color in the same places.

Fish or Flatworm?

Some cleaners change their occupation when they get older, and other fish as well may vary their life style from youth to old age. Young fish are especially in danger of being eaten and may use various "disguises" to avoid predators while small. Some juvenile fish are camouflaged as floating bits of dead leaves and grow up to look completely different. Others may masquerade as completely different animals.

Scientists collecting fish on a coral reef in the South Pacific once saw what they thought was a flatworm swimming near the bottom. It was black with a bright orange border, a typical sort of flatworm color pattern, and it swam just like a flatworm. At first they ignored it, for flat-

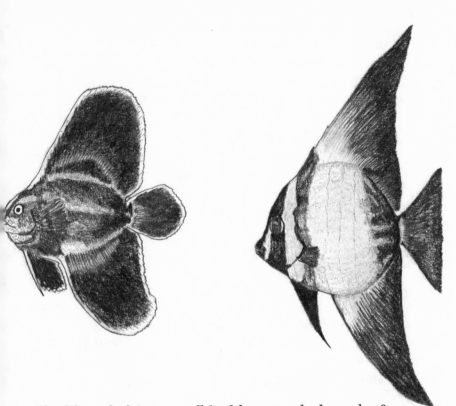

The fish on the left is a small batfish, captured when only 16 millimeters long. It resembles a marine flatworm, most of which are poisonous and left alone by experienced predators. When it grows further, right, it takes on its adult appearance and changes its swimming position. Drawing by the author.

worms generally fall apart when collected and pickled. But then they looked closer and saw that what looked like the top of a flatworm was really the side of a fish. They could distinguish the fish's eye and the side of its mouth. And they realized that the "body" of the worm was really made up largely of a fish's fins. They captured this strange little fish and found that it was a young batfish. Adult batfish look very much like the familiar freshwater angelfish.

They are white with black bands and have no trace of orange on their bodies. While the young fish have broad fins with rounded borders which give them the flatworm-like outline, the adults have long, more pointed fins. Marine flatworms tend to be poisonous and are rarely eaten. With its flatworm disguise, young batfish avoid being eaten while small. As they grow, they lose their black and orange color and begin to swim upright in normal fish fashion.

A Mimic of the Moray

One of the most well armed of reef fishes is the moray eel. Its pointed jaws are armed with impressive rows of very sharp teeth, and it will attack without hesitation. The appearance of the moray is unmistakable, with its well armed mouth and glaring, prominent eyes. Morays live in reef crevices and, even if startled by a larger fish or a scuba diver, leave their threatening heads exposed while keeping their tail ends protected within crevices.

One kind of moray is especially striking in appearance, with a dark body marked by small white or bluish dots. Living within the same geographical area as this moray is an uncommon fish from a completely different family which has the same sort of coloration as the moray. While most reef fish will do their best to disappear among the cracks and holes of the reef if threatened, this fish behaves quite differently. It darts quickly toward the reef, but hides only its head, leaving its rear end completely exposed. The fish spreads out its tail fin and top fin, revealing a striking if small eyespot. The rounded tail fin is

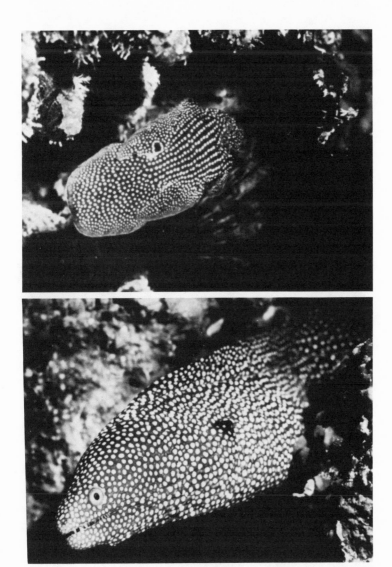

D. POWELL, STEINHART AQUARIUM

By partly entering a crevice in the rocks, the fish Calloplesiops
altivelis *leaves its eyespotted rear end exposed and so presents
to predators a threatening mimic. The spotted skin and eyespot
together effectively suggest a dangerous moray eel, below.*

shaped like the head of a moray, while the eyespot is in just the right spot to mimic a moray's eye. The rear end of this harmless hider now looks for all the world like the head end of a vicious moray. While a predator may hang around waiting for fish to come out of hiding if they simply disappear, a predator startled by the moray-mimic display may flee quickly, allowing the mimic fish to resume its normal business more quickly than if it had merely hidden as other reef fish do.

The Poisonous Red Eft

Most people know that there are animals called salamanders and newts, but how many have actually seen a wild one? These relatives of frogs and toads usually keep out of sight under logs and stones in woods or are concealed among the debris at the bottoms of ponds. But now and then a hiker is startled when a small and bright red-orange salamander marches across his path in broad daylight. This is very unusual behavior for such ordinarily secretive creatures, and one could hardly miss seeing such a brightly colored animal on the brown forest floor.

The red eft is a young stage of the red-spotted newt. Adult red-spotted newts are dull green creatures with a row of red spots along each side. They live in ponds in the eastern part of the United States. Their eggs are laid in the water and hatch into larvas with gills. In some areas, these larvas grow and develop into dull green newts just like their parents and never leave the water. But in other areas, the newt larvas undergo a striking change. As they grow, they change color, becoming bright orange or red-orange. They leave the water and march out onto land as red efts. A red eft may spend one, two, or even three years on land before it returns to the water and changes into a dull green, water-bound adult.

Although scientists are not sure why the red-spotted newt has the unique red eft stage in its life cycle, they do know that the bright colors of the eft are a warning to enemies to stay away. The skin of the red eft contains potent poisons which cause muscular weakness, gasping, and convulsions as well as vomiting. They apparently taste bad, too, for if mice are force-fed red efts, they throw up and rub their mouths on the ground as if trying to get rid of a bad taste.

Another salamander found within the range of the red-spotted newt is called the red salamander. There are a couple of species of red salamander, and each has more than one form. In some areas, red salamanders are not really red, but rather are brownish pink or dark reddish brown, becoming darker as they grow older. In other places, red salamanders are truly red or orange as young animals, but turn dark as they age. Still others remain red throughout their lives. Why should there be such variation in the colors of the different red salamander populations?

Scientists studying this problem have decided that the red salamander is a Batesian mimic of the red eft. Its red color is similar to that of the red eft, and both species also have spots. But because the mimicry is less exact than that found in most kinds of insects (which offer the most numerous and well known examples) many scientists have doubted that the red salamander is really a mimic of the red eft. The red salamander is usually larger than the eft, and its color is actually brighter. While the spots on the eft occur in rows along its sides and are bright with black rings around them, the spots of the red salamander are found all over the back and are black.

The only good way to resolve a scientific argument is to do some experiments. When that was finally done with red efts and red salamanders, the case for mimicry was really strengthened. Birds with no experience would eat red salamanders without hesitation. But after being presented with a red eft and tasting it, 16 of the 17 birds used in the experiments refused to touch either species. Because biting into a red eft is such a very unpleasant experience, birds are especially wary of repeating it. As we said earlier, the more obnoxious the model, the less exact the mimic need be to gain protection. And predators are more likely to encounter efts than red salamanders, since the efts are quite common and active during the day, while red salamanders are usually hidden and come out at night. Still, predators which are familiar with red efts may find red salamanders while hunting for food in the leaf litter of the forest. Presumably, when they do so, they leave them alone.

Further evidence that red salamanders have evolved as mimics of efts is found in their behavior. When bothered, most edible salamanders try to escape. But a disturbed red salamander holds its ground, waving its tail to the side or lifting it as straight up as it can. This display mimics quite closely the display of red efts when disturbed. Efts, too, wave their tails to the side or lift them when bothered. In the case of the eft, the display "reminds" the predator that this is no ordinary salamander; in the case of the red salamander, the display increases its resemblance to the eft.

The color-form distribution of red salamanders is the final bit of evidence that mimicry is involved here, for in areas where the "red" salamander never has a truly red stage, red-spotted salamanders also lack a red eft stage. Varieties of red salamander which are red only when young get much larger than efts as adults, whereas types that remain small are the ones which are red their whole lives. Thus, the color patterns of the red salamanders appear to be influenced by the distribution and size of the red eft. This is certainly strong evidence that the evolution of the red salamander has been strongly influenced by the protection gained from its resemblance to the red eft.

The Coral Snake Mystery

Coral snakes are small but often deadly poisonous snakes belonging to the same family as cobras, kraits, and the most poisonous snake of all, the mamba. They have bright rings of red, yellow, black, and sometimes white, which stand out strikingly. Many species of snakes from another family (the Colubridae) also have the same type of pattern. Sometimes species from the two families look so similar that they must be placed side by side before they can be told apart. Some of the colubrids with the coral-snake pattern are poisonous, while others are harmless. At first glance this looks like a typical mimicry ring, with varyingly dangerous species being copied by harmless ones.

But let's wait a minute; there is a serious problem here.

Some coral snakes are so poisonous that they kill any creature they bite. If a snake-eating predator samples such a coral snake, the predator dies. And a dead animal can hardly learn to avoid warning coloration. If mimicry is involved here, it must be quite different from the kinds of mimicry which we have discussed so far. And perhaps these snakes share the same sort of color pattern for a completely different reason which has nothing to do with mimicry at all. Within recent years scientists have argued for all of these possibilities; let us look at the arguments one by one.

Many animals which look conspicuously colored to us are really camouflaged in their natural surroundings. A spotted leopard or cheetah may catch one's eye in a zoo, but out in the dappled shade at the forest's edge, these cats are very hard to see. Many small marine animals, such as snails, have conspicuous colors which stand out in a plain aquarium but blend in perfectly with the bright colors of the coral reefs where they live. Some scientists argue that poisonous coral snakes and their "mimics" are actually sharing an effective camouflage design rather than a conspicuous warning one. They say that, in the typical coral-snake habitat of dappled sun on the forest floor, such a contrasting pattern of bands can break up the outline of the snake's body and make it hard to see rather than obvious. But others point out that coral snakes live in many different habitats other than forests, such as mountains and deserts. In these environments, the snakes are obvious to the eye. So camouflage cannot completely explain coral-snake rings.

Other scientists stick to the notion of classical warning

coloration and mimicry. They point out that, among social predators anyway, one animal could observe the effects of a coral-snake bite on another of its kind and learn to avoid such animals in that way. But since most predators are solitary animals and are unlikely to stumble across an encounter between one of their kind and a deadly coral snake, this explanation is rather weak.

Wolfgang Wickler, an expert on mimicry, has come up with an ingenious explanation for mimicry in coral snakes which he calls "Mertensian mimicry," after the great snake expert Robert Mertens who studied the coral-snake problem in detail. Wickler agrees that the deadly poisonous snakes cannot be the models in these mimicry rings, for any predators which sampled them would not live to learn the warning pattern.

He states, rather, that the mildly poisonous snakes are the models. A predator which attacked one of them would have a nasty experience but would survive to remember and avoid the coral-snake pattern. In this scheme, the harmless snakes would be Batesian mimics of the mildly poisonous ones. The deadly coral snakes would be the Mertensian mimics. They cannot be called Müllerian mimics, since their pattern does not help predators to learn the common pattern. In fact, the predators are protected from death by confusing the deadly poisonous snakes with mildly poisonous ones. Thus, in Mertensian mimicry, the most dangerous species in the mimicry complex is not a model, because it is too dangerous to allow the predator to learn. The less dangerous species are the models, with both the harmless Batesian mimics and the very dangerous Mertensian mimics copying them.

Clever as this interpretation may be, there are serious problems in it. While it can theoretically explain the resemblance of the mildly and deadly poisonous coral snakes to one another, it does not explain coral snakes in all areas. In parts of the southern United States and Mexico, highly venomous and completely harmless coral snakes are found together without any moderately poisonous ones with similar patterns to serve as models.

And it cannot explain another case of snake mimicry in southeast Asia, where a deadly poisonous coral snake and a completely harmless colubrid bear a striking resemblance to one another. Not only do these two snakes look alike, they share the same behavior pattern when disturbed. They roll over on their backs, revealing a pattern of black and white bands on their bellies. Both also have a bright red underside to the tail which is raised and displayed presumably as a false head when the animal is under attack. It would be difficult to argue that these two snakes are not copying the same physical appearance and behavior, yet there is no similar moderately poisonous snake around which could serve as a model for them. Therefore, one can only conclude that the poisonous snake is the model, which brings us right back where we started from.

There is still one way out of this dilemma. Suppose that evolution acted on the behavior of snake predators. An individual which went right out and attacked a deadly snake would die without leaving any offspring. But if another individual predator had an inherited tendency to avoid the brightly banded pattern, it would survive to reproduce and pass on this trait to its offspring. In theory, then, it seems possible that evolution could act on the behavior of predators, eliminating individuals which attacked the coral-patterned snakes and preserving those with an inherited avoidance of them.

Up to now, only a few preliminary experiments have been done to investigate the intriguing question of whether an avoidance of warning color patterns can be inherited. Dr. Susan Smith has done experiments using birds called great kiskadees. These birds live in fairly open habitats all the way from Texas to Argentina. They are known to eat small reptiles, and many kinds of coral snakes, including deadly ones, are found within their range.

Dr. Smith took young kiskadees from their nests before they could have had any experience with coral snakes and raised them in the laboratory. She then took rounded sticks of wood and painted them with various patterns. She used ring patterns, which circled the sticks like the rings on a snake's body. The color patterns she used were green and white, yellow and red, and a coral-snake pattern of wide red, narrow yellow, wide black, and narrow yellow rings. She also painted some with yellow and red stripes and others with coral-snake stripes which ran from one end of the sticks to the other instead of circling them. When Dr. Smith placed sticks painted with solid white, green, red, yellow, or black in their cages, the young birds attacked them immediately. They also attacked the sticks with green and white rings and those with yellow and red stripes.

But when she placed a stick with a coral-snake pattern in the cage, no bird would approach it. Four of the birds even let out a loud alarm call when they saw the coral-snake-patterned stick. The birds also avoided the sticks with red and yellow rings on them and approached the coral-snake stripes very carefully. Since these birds had no chance to learn that the coral-snake pattern is dangerous, Dr. Smith came to the conclusion that such avoidance can be inherited. If this is indeed the case, Mertensian mimicry is an unnecessary concept. A modified form of Müllerian mimicry would be operating instead, with evolution influencing the behavior of snake predators as well as the patterns of the snakes. Predators with an inherited tendency to avoid coral-snake colors would not attack the poisonous snakes and would have a greater chance of survival than predators with no native fear. The highly poisonous and moderately poisonous snakes would serve as Müllerian models for one another, and the non-poisonous species would be Batesian mimics of both the deadly poisonous and moderately poisonous kinds. One hopes that more research will be carried out on the behavior of predators towards coral snakes so that we can find out if inherited avoidance of warning colors is a general phenomenon or not. Because we know so little about the inheritance of behavior, many scientists are doubtful that natural tendencies such as avoidance of certain color patterns can be caused by heredity rather than by experience.

6 · Mimicry of Social Insects

People have always been especially fascinated by the social insects. Insect societies appear to mirror human societies in some ways, with cooperation, division of labor, and a system of different "social classes." But the differences between insect and human social structures are really much greater than the similarities. Remove an ant or termite from its colony and it will die, for it is incapable of existing alone for long. While people are able to think and can be flexible about what role they play in society, social insects cannot think or choose their places; they are determined by biology.

One striking area of difference between social insects and people which most people know nothing about is the existence of a vast variety of insects and spiders which live in or around ant and termite colonies. Hundreds of species of spiders, beetles, and flies as well as silverfish and other insects may be found associated with these social insects. Some, such as the aphids which many ant species care for, could be compared with our domesticated animals. The ants take care of the aphids, and the aphids provide food for the ants. But most of the "ant guests" and "termite guests" are uninvited. They are just there, gaining food and/or protection from their hosts while giving little or nothing in return.

While a great variety of insects which live with ants and termites have been discovered, very little is known about their ways of life. Some are found deep within their host's nests and never emerge into the sunlight. Others accompanying army ants or driver ants on their raids. Some of these creatures are only casual associates of social insects, while others cannot survive without them. Because social insects live in large colonies and since most kinds cannot live for long in a laboratory, it has been very difficult to study them or their associates.

Look-Alike Spiders

Many of the spiders found with ants are clearly visual mimics. But it takes a few tricks to turn a spider body into an antlike one. While ants have three obvious body regions, spiders have only two. Ants have antennas and three pairs of legs. Spiders have no antennas but possess four pairs of legs. Instead of antennas spiders have a pair of appendages called pedipalps. Most spiders hold their pedipalps tucked close to their mouths. Some ant-mimicking spiders, however, create the illusion of a three-part body by holding their pedipalps out in front of their bodies. Their pedipalps have enlarged ends. When the pedipalps are held together in front of the spider, the enlarged ends give it a false "head."

Other spiders which mimic ants have a constriction in

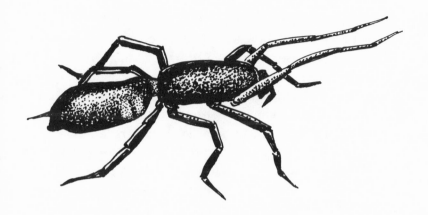

This is Micaria longipes; *not only does the body of this spider resemble that of an ant, but its gait is antlike and it holds out its two front legs as if they were antennae.*

the middle of the front part of their bodies which makes it look like the separate head and thorax of an ant. Ant-mimicking spiders often hold their front legs up next to their heads and wave them about like antennas. The resemblance created by these changes are often very striking and are reinforced by the animal's behavior. While most spiders spend much of their time sitting completely still, ant-mimicking spiders are mostly on the move, walking with a quick, nervous, antlike step.

Who is being "fooled" by these antlike spiders? Do the ants view the spider as another of their kind and therefore ignore them, getting close enough to be eaten? Or are predators confused into thinking the spiders are really distasteful ants? Many ants are rarely eaten by birds and other hunters because of their defensive acid secretion or their nasty bite. Many ant-mimicking spiders resemble ants which are especially well-equipped to defend themselves. The ants run fearlessly in the open and the spiders join them. These spiders may also be "hiding" from predators which concentrate on eating spiders, such as certain wasps and hummingbirds. By looking non-spiderlike, they avoid being eaten by these specialized predators.

The best-known ant-mimicking spiders are jumping spiders from one particular group. The red weaver ants of Africa and India are often associated with a red spider which looks like them. Black ants, however, are accompanied by a related black mimic. The black spiders are never found with red ants and red spiders are never associated with black ants. Spider-hunting wasps appear to be fooled, for very few of the ant-mimicking spiders are found in wasps' nests, even though they may be very common in some areas. The ants themselves, however, can tell the difference, and the spiders keep well away from main ant runways.

The red spiders associated with the weaver ants have a serious problem, for these voracious ants will attack and eat just about any living thing. But their protection from spider-hunting wasps is greater. While black-ant mimics are sometimes recognized as spiders by wasps and taken from among the ants, red ones never are. Even if the wasps can recognize the red spiders, they leave them alone. Weaver ants are powerful enough to kill wasps, and getting too close to them would be foolish.

Ant mimicry by spiders can get quite complicated. One ant-mimicking spider found in Costa Rica copies different ants at different life stages. These spiders do not associate directly with ants, but live among the dead leaves of the forest floor, as do many kinds of ants. The very young, small spiders are black and shiny and can easily be mistaken for various small ant species. When they get

The habit of some ant-mimicking spiders of holding up their front legs may start in such a way as the action does with the wolf spider—not considered a mimic—seen here. The spiny legs may help to locate prey or a mate, but this has not been proved. If some members of this species gradually evolved bodies that looked more and more antlike, they would become ant mimics and the raised legs would help out the illusion of an insect—six legs, and antennae poised in front.

DR. PHILIP CALLAHAN

older, the ants become a yellow-orange color, like some medium-sized ants. Adult female spiders are a reddish brown or almost black color like several kinds of large ants, while males are bright orange-red and are more specific mimics of particular ant species. At all life stages, the spiders walk around with a jerky gait, with their front legs held forward and waving, like ant antennas.

One ant-mimicking spider from Asia and Australia does not look much like the ant it associates with, and its movements are not strikingly antlike. This spider feeds on ants and apparently lures them by stationing itself near an ant column and making feeble movements like those of an ant in trouble. Attracted by what seems to be a nestmate in need of help, an approaching ant is pounced upon and sucked dry. Because the ants have poor eyesight, these spiders need only a general resemblance to ant behavior to "fool" their prey.

Protected Young Insects

The spiders which "copy" ants and so avoid being eaten do not actually live inside ant colonies. While some kinds do join ant columns on the march, most merely are found in the same type of habitat and are presumably easily mistaken for ants by birds. Insects which share ant habitats often have young stages which mimic ants, presumably also avoiding being eaten. Young lobster moth larvas do not look at all like typical caterpillars. Their bodies are thin, and their front false legs are long and slender. They wave these in front of their antlike heads like antennas. While most caterpillars live singly, lobster moth caterpillars tend to cluster on small branches like ants looking for aphids with honeydew.

In Africa, ants are found on just about every tree and shrub. Other successful insects often have complex relationships with them. Mantises live on plants, too, and often eat ants. But ants may eat unprotected mantis eggs, and females of several kinds of African mantises guard their eggs. When the eggs hatch, the emerging mantises are very antlike in appearance and may be hard to distinguish from the ants. This resemblance may prevent ants from bothering them as well as protect them from hungry birds. Some kinds have a general resemblance to black ants, while others may mimic very closely the precise coloration and markings of particular ant species. When the mantises become larger and can no longer "pass" for ants, they lose their antlike look and become camouflaged.

Ant Specialists

One particular family of beetles has many members which are associated with ants. Scientists estimate that life with ants has evolved at least 20 separate times in this one family. Why should this be so? The family is huge, with over 28,000 known species. Most of these beetles live in the rotting debris of the soil and feed on insect larvas and other small animals. Since ant colonies are also found in the same kinds of environments, these kinds of beetles would often come in contact with ants. Since they feed on small, soft-bodied creatures, ant larvas make a perfect food for them. Any such insects which developed some way of being left alone by ants would have a big survival advan-

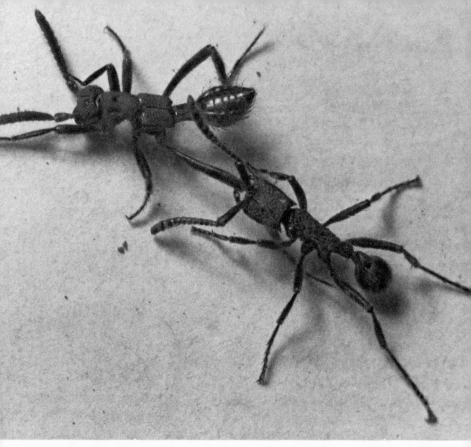

The beetle at left, Ecitosius robustus, *mimics surprisingly well the Costa Rican army ant at right.*

tage, since they would then have easy access to food and protection.

Beetles associated with ants sometimes merely accompany their hosts on raids and snatch away some of the booty. These beetles do not appear to have become ant mimics in any way. Others are obvious visual mimics, while some are so completely integrated into the ant colonies that they probably mimic the very odors of the ants themselves.

Because their colonies live out in the open, making them easy to observe, army ants are among the best-studied ants in the world. The beetles which accompany them are also better known than others which live with ants. There are at least 210 species of beetles in the one family alone which are associated with army ants, and beetles from other families may be present also. Some of these merely "hang around" on the edges of the colony, but others are completely integrated into it.

Some beetles associated with army ants look so much like the ants that they are hard to distinguish from them as the ant columns march over the forest floor. Beetles have short, stocky bodies, while ants have long, thin ones. Therefore the modifications necessary to get from a beetle type of body to an antlike one are considerable. One key feature of the ant body is the petiole, a narrow constriction between the thorax and abdomen. Ant-mimicking beetles have evolved at least seven different ways of copying the petiole.

Because so many of these beetles have developed some kind of petiole, some scientists who study them think that the narrow "waist" is an especially important part of the beetles' "disguise." They have observed ants meeting along the trail stopping to examine one another with their antennas. When they reach the petiole, they pause for a moment, then separate and go along their way. Ants encountering the mimetic beetles go through the same ritual with them, departing after feeling the petiole. Other scientists think that the petiole is simply an especially important part of an antlike appearance. Since army ants are blind, a visual mimicry of the ants would hardly help the beetles to be integrated into the colonies. These scientists feel that the visual mimicry fools predators, and that odor mimicry is more important in convincing the ants to accept the beetles. Very few animals eat army ants, so an antlike beetle which lives in the midst of an army ant colony is well protected against being eaten, as long as it can fool the ants as well.

Ants rely on their sense of smell to tell them about the world around them. They also use it to recognize friends and enemies. To be successful in living with ants, beetles must develop either the same smell as the ants or an acceptable alternative. Unfortunately we know very little about this aspect of ant-mimic life. Those beetles which look most like ants and live most closely with them are also the beetles which rarely survive more than a few hours after arriving in a laboratory, so observations on them are limited. During their few hours of life in captivity, however, many of these beetles spend most of their time grooming their hosts. Grooming behavior is common in many social animals. One animal strokes another or two animals stroke each other. Ants do a lot of grooming among themselves, and accept grooming from a beetle just as they would from one another. Probably during this intensive grooming the beetles take on the scent of the ants, allowing them to be accepted into the colony.

Meanwhile the beetles are busily taking advantage of their hosts. Rather than wasting energy marching along in the ant columns, many of the beetles ride on top of cocoons or booty carried by the hard-working ants. Some are actually carried in the ants' jaws like larvas. While the ants are at rest, the beetles busy themselves eating ant larvas or hard-won booty from raids. When an ant is full of food, it will share some with its nestmates, and some ant "guests" can get the ants to feed them as if they were ants themselves. The hungry guest taps the ant's body with its antennas or forelegs. When the ant turns to face it, the "guest" taps it repeatedly on its mouthparts, just as a hungry nestmate would. The deluded ant gives up a drop of food, which is then devoured.

Some beetles leave their offspring in the care of ants. The ants tend them just as they do their own brood. Two kinds which scientists have observed are especially well adapted to life among the ants. The beetle larvas are bigger than the ant larvas, but the ants do not seem to notice. The beetle larvas beg with motions similar to those of the ant larvas, but are more intense about it. They are well fed by the ants. The beetle larvas also produce chemicals in small glands all along their bodies which "fool" the ants into tending them. If these attractants are extracted from the beetle larvas and applied to dummies, the dummies are treated as if they were living ant brood.

Termite "Guests"

Termite colonies, too, are exploited by many kinds of what we might call guests. Beetles again lead the list, but flies and other insects live there as well. When we think of termites, we have in mind the dangerous little house-eater common in this country. But termites in the tropics are quite different. They live in large underground nests. Often part of the nest extends above the ground. Such termite nests are several feet tall, with walls as hard as cement; and these have millions of inhabitants. As with ant guests, all degrees of integration into the colony can be seen with termite guests.

Since termites live underground where it is dark, we would not expect their guests to look like them, and indeed, they do not. But strangely enough, the most specialized termite guests do look very much like one another, even when one is a fly and the other a beetle. Termite guests often have enlarged abdomens, and many carry their abdomens curved up over their backs so that the tip is above the insect's head. Although it is still a mystery why their abdomens should be so enlarged, we do have some clues as to why this body form should be so successful.

Like ants, termites live in a world of important odors. Scent is their most vital sense, and they do a great deal of mutual grooming. The gland system of one beetle living with termites is very suggestive. This beetle is a typical termite guest, with a small head and thorax and an enlarged abdomen carried over its head. In addition to the glands found in free-living beetles, this species has a large gland which opens near the tip of its abdomen and makes secretions which are licked by the termites, as do other glands located along the sides of its abdomen.

Unfortunately, scientists have yet to identify these substances chemically. They could simply be a source of nourishment for the termites, in which case the termites would be gaining at least something from their guests. The secretions could be some sort of attractive or drugging substance which makes the beetle acceptable or destroys the termites' aggressive tendencies. Or finally, the chemicals could be identical or very similar to the termites' own body secretions, so that the beetles are truly mimicking their hosts. Interesting as guests are, only if the substances they produce copy their host's secretion could these social parasites be considered odor mimics.

Beetles living with termites may also use tactile mimicry, however. One kind of beetle living with a South African termite has a larva which may mimic termite nymphs. Its head and thorax together look like the head of a termite nymph, and termite workers react to it just as if it were one of their own. Another beetle living with the same kind of termites holds its abdomen over its head just like so many termite guests. But its abdomen is divided into three obvious regions which give the impression of the head, thorax, and abdomen of a young termite. Scientists have been fascinated by these beetles but unfortunately they are quite rare; fewer than 400 have been found, even though collectors have hunted and hunted for them.

The most striking of all these possible tactile mimics is even rarer; only one specimen has ever been located. It, too, was found in southern Africa. The abdomen of this beetle is incredibly swollen. It is carried over the animal's head and is marked with various lumps and bumps which make it look very much like a worker termite. Not only that, but there are also strange appendages hanging from the abdomen which give the impression of a pair of antennas and three pairs of legs. Thus the abdomen of this remarkable creature may function by itself as a mimic. Since the nest is dark and the termites are blind, the effect of visual mimicry which we see must be the consequence of a tactile mimicry which would "fool" the termites.

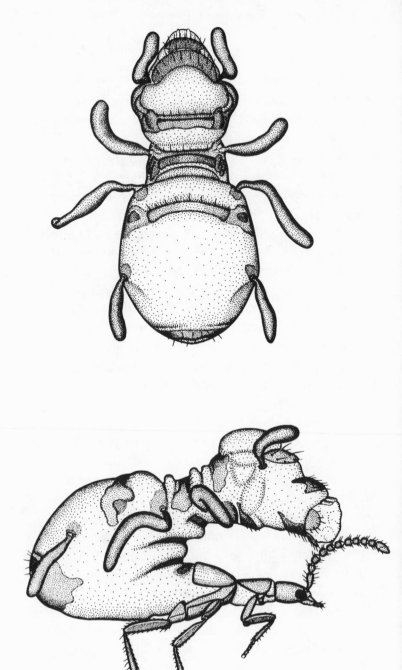

A Bee or Not a Bee?

For over 2000 years, people thought that a swarm of bees could be produced from the rotting carcass of a bull or ox. Complicated directions were written by many authors on just the right way to kill the bull and prepare the body so that honeybees would result. According to the great Roman poet Virgil, the bull must be very carefully chosen in the springtime, under the astrological sign of Taurus, the bull. The poor creature had to be clubbed to death or suffocated, not bled in any way. Finally, in the eighteenth century, a scientist pointed out that the "bees" which magically arose from decomposing bodies were not bees at all but were flies which bore a striking resemblance to worker honeybees. The female flies lay their eggs in rotting flesh, and the maggots feed on it. Not only were there no bees, there was no magic.

These honeybee mimics are called droneflies. Their behavior as well as their appearance is honeybee-like, for they buzz like bees and feed on flowers like bees. The Browers experimented with feeding honeybees and droneflies to toads. They showed that if a toad got stung in the process of eating a honeybee, it was just as cautious about the flies as about bees from then on. If the wings of either the bees or the flies were cut off, reducing the buzzing

This remarkable beetle, Coatonachthodes ovambolandicus, *is the best mimic of termites, states Dr. David Kistner, its discoverer. It even has clublike appendages that mimic antennae and legs.*   DR. DAVID H. KISTNER

sound, the toads were more likely to eat either one. Thus the sound mimicry of the droneflies seems to be an important part of their "copying."

Many kinds of flies mimic bumblebees, too. The large narcissus bulb fly has dozens of color forms, all of which mimic different kinds of bumblebees. Many of its hoverfly relatives do the same. Robberflies are another group of bumblebee mimics. The Browers found that bumblebees gave toads a more painful sting than the smaller honeybees, so toads were much more reluctant to attack them after being stung. This protection was also extended to their robberfly mimics. While experienced toads rejected 30 to 57 per cent of the droneflies offered, they refused 93 per cent of the robberflies.

Not only do robberflies resemble bumblebees, they eat them. They are effective Batesian mimics of the bees, but their similar appearance may also help them get closer to their prey. This would make them aggressive as well as Batesian mimics. While this may be partially true, it looks as if the robberflies hide as much as possible from their intended prey. A hungry robberfly will perch on the shady side of a plant stem near flowers which bumblebees visit. The fly approaches its prey from above and behind, grabs the bee in midair and stabs it with its sharp mouth, rapidly injecting a deadly poison. The process is so fast that the bee has no time to react. Since these flies also prey on honeybees, wasps, and other insects, their resemblance to bumblebees is probably mostly protective.

The cuckoo bees present another confusing case of bee mimicry. Cuckoo bees are closely related to bumblebees, and this may be why they look so much like them. But their close resemblance may also make it easier for them to invade bumblebee nests. Each kind of cuckoo bee parasitizes a particular kind of bumblebee which it closely resembles. The cuckoo bee queen enters the bumblebee nest and takes over, either killing the bumblebee queen or destroying her eggs if she manages to lay any. All the young bees produced are cuckoo bees, and the bumblebee workers do the work of raising them. Since some kinds of flies which are either predators or parasites of bees also resemble their hosts, it seems that the resemblance of a bee predator or parasite to its intended victims is probably of some advantage. So little scientific study has been done on these interesting insects that no conclusions can be drawn as yet.

Other insects mimic bees, too, including some beetles. The markings on the top of a beetle which one sees when examining it are really on the hard front pair of wings, not on its body. When beetles are not flying, the front wings are held against their bodies. When they fly, the wings are spread out. Therefore a beetle which has black and yellow markings on its front wings and resembles a bee when sitting on a flower loses its beelike look when flying. At least one bee-mimicking beetle has gotten around this problem with a modification of its front wings. While some of its relatives with bee markings fly in the normal way, this beetle cannot. Its front wings are inseparably joined together over its back by a firm tongue-and-groove joint. When it flies, the front wings remain in the resting position close to its body, while the hind wings are used to fly. In this way, the beetle always keeps its disguise, whether flying or at rest.

7 · Plants Do It Too

With a few exceptions, mimicry in plants is a neglected topic. Perhaps there are few plant mimics, or perhaps scientists just have not looked in the right places. We already have seen how the gold-of-pleasure, a useless weed, mimics the right characteristics of flax to be accidentally cultivated by humans. While this weed is an annoying pest, two other plants which started out as weeds have become valuable crops.

Wheat and rye both began as wild grasses. They both had small seeds which broke off the stalk when ripe. These are valuable traits for a wild grass. Small seeds are perfectly adequate for producing new plants which grow as fast as grasses do, and seeds which break off easily are spread about easily. But different qualities are desired in cultivated grains. Humans value traits that are the opposite of those needed in the wild. Large seeds are much more useful as food. Seeds which break off easily become lost when the grain is harvested and transported, so cultivated grains need the kind that remain attached.

When people began to grow wheat as a crop, they

selected and replanted large seeds which remained on the stalks. Wild rye was also growing in the fields with the wheat, and it also came under the same new selective pressures as the wheat. Rye which continued to grow as a weed in the wheat fields also had bigger seeds which fell off less easily. For many generations, rye grew quietly and unnoticed in the wheat fields. Through unconscious selection by humans, it evolved from a long-lived weed with small seeds which fell off easily into an annual plant with useful large seeds. Eventually, this "new" plant was noticed, for it could grow under less favorable conditions than wheat. In areas of poor soil, rye grew where wheat failed. Rye had already taken on the desirable traits of a crop plant by the time it was noticed, and it was readily developed as a crop in its own right.

Oats came into use as a crop in much the same way. While wild oats have coarse ears and seeds which readily break off, the oats which grew as weeds in wheat and barley fields developed in the same direction as rye weeds. Since oats will grow in very poor soil where wheat and barley fail, eventually the value of this plant was recognized, and it was also grown as a crop.

Fruitlike Seeds

Plants cannot move around as animals can, so they need a way to distribute their seeds. If all the seeds of a plant simply fell around it, the new plants would be crowded together and most could not survive. Some plants such as dandelions have "parachutes" which allow the wind to carry them to new places. But the most common way for plants to disperse their seeds is to get animals to do it for them. Many weeds living in grasslands have sticky or spiny seed coverings which cling to animal fur. A tremendous variety of plants "hide" their seeds inside tasty fruits. Birds or other animals eat the fruit and the seeds are deposited unharmed in new localities in their droppings.

Producing nutritious fruits, however, takes energy. It would be much more efficient if plants had a way of tricking birds into eating their seeds whole without giving the bird any reward. A few plants appear to have evolved ways of doing this. They have seeds which resemble berries, and birds are thought to eat them as if they were berries. Such plants cannot be very common, or birds would soon learn to ignore them in favor of more fruitful eating.

One such plant is a wild onion found in North America. It lives in woods where many kinds of berries grow close to the ground. The onion's seeds are round, hard, and shiny black. They look like tasty dark berries, and presumably birds hunting for berries will eat the onion seeds as well.

Many plants in the family to which peas and beans belong have seeds which appear to mimic berries. Some have shiny bluish seeds, but most have red or red and black seeds. These are apparently mimicking two different kinds of fruits or berries. Berries which are familiar to us have juicy flesh covering the whole seed or seeds. The red mimetic seeds resemble these. But in tropical regions, some plants have red fruits which only partially cover the shiny, contrastingly colored seeds. The black and red seeds appear to "copy" these.

Few actual scientific studies have been done on mimetic seeds. One scientist found that seed-eating birds rejected berry-like seeds of one type, but fruit-eating birds ate them. They emerged unharmed in the birds' droppings. Red and black mimetic seeds are found only in tropical regions, just like their supposed model fruits. Red mimetic seeds have a much greater distribution, just as one would expect if they were "copying" red berries. Although records do not exist saying how common all these plants are, many are known from only a single specimen. This would indicate that they are rare, as they should be if they are mimics.

"Fooling" Flies

Some scientists think that plants that become pollinated by "deceiving" insects are not mimics in the strict sense of the word. But others argue that a flower which smells like rotting meat or one which looks like an insect should surely be considered mimics if they succeed in attracting insects in this way. Plants need help to fertilize their flowers. The pollen, carrying the male cells, must somehow be carried to the ovaries of a flower which contains the eggs. Many plants, such as pines and grasses, simply produce large amounts of pollen which the wind carries to the female flowers. Others attract insects, bats, or hummingbirds with white or brightly colored flowers and inviting aromas, giving them sweet, nourishing nectar in trade for their services in spreading pollen about.

But a few flowers use flies as pollinators. Since flies deposit their eggs in rotting meat, flowery scents will not

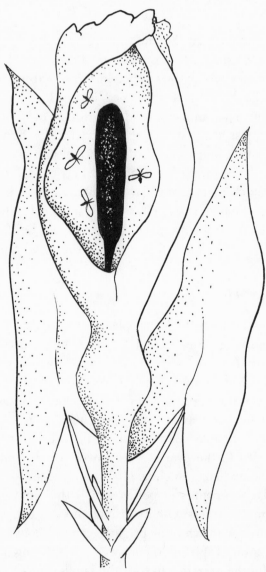

The wake-robin, related to our jack-in-the-pulpit, gives off a disagreeable smell that attracts flying insects. It traps them temporarily, during which time they pollinate the plant.

do for these plants. Instead they smell like rotting meat. The starfish flowers of South Africa are particularly "unfair" in their dealings with flies. Their large, hairy flowers smell foul to humans and are the color of decaying meat. Flies are so completely deceived by the starfish flowers that females actually lay their eggs inside them. The maggots die after hatching, for the flower has nothing to offer them in the way of food. But the adult flies, in their journeys from one flower to another, do the job of pollination.

Many other flowers attract flies, beetles, or gnats with an odor similar to rotting meat. The Dutchman's pipe is one. This common vine has long, thin flowers. When they open, the flowers point upward. Their strange odor attracts gnats, which slip and fall into the flower when they land. The inside of the flower is covered with tiny grains of wax which come off on the insects' feet, making it impossible for them to maintain a grip. Also lining the flower are hairs which point downward. The hairs can be bent further down, but have a thick base which prevents them from being pushed up. Once the gnats have slipped into the flower, they are trapped. But at least the Dutchman's pipe provides nectar for its captives. While they feed, they rub off any pollen from their bodies which they acquired from previous flowers and pollinate the flower. The gnats are trapped for two or three days, during which time they become dusted by pollen. Gradually the flower bends over and the hairs inside wilt, allowing the gnats to escape and go on to another Dutchman's pipe.

Some kinds of orchids are also pollinated by flies. Various kinds of "deception" attract the insects. Some orchids have furry warts or spots which lure flies, possibly because they look like other flies which have already landed. And many fly-pollinated orchids emit rotten odors of various kinds, like those of decaying meat, vegetation, or fruit.

Orchids Mimic Insects

A very different way of insuring pollination is used by many orchids. Part of the flower so resembles a female insect that males try to mate with it, pollinating the flower in the process. The common Mediterranean ophrys orchids resemble various insects, and their common names show that people can also see the resemblance; one ophrys is called the bee-orchis, another the sawfly-orchis, and still another the fly-orchis. The flowers of many ophrys do indeed have a startling resemblance to an insect. Shiny spots resemble the eyes of a bee, and metallic blue mirror-like areas resemble marks on the female bees' bodies.

The center of an orchid found in the Andes looks remarkably like the body of a female fly. It has yellow and reddish brown stripes like the body of the fly, and has parts stretching out to the sides like wings. The tip of the "abdomen" reflects the sun, like that of a female fly ready to mate. When a male fly comes along, it stops for a moment to investigate the flower, then leaves. But a short visit is all that the flower needs, for in the process of touching the flower, the male fly's abdomen picks up pollen. When he visits another flower in his search for a mate, he transfers the pollen to it.

Other orchids are different. Male bees of some kinds set up territories which they defend against all comers.

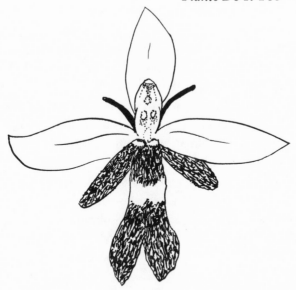

This orchid mimics a female bee, thus attracting a male bee that investigates it and incidentally brings about pollination. It is one of several orchid species that have developed mimicry.

They will attack any other male of their kind which invades their "property." The flowers of a kind of orchid found in Ecuador, where some of these bees live, have very slender stalks. The gentlest breeze sets the flowers to dancing. To human eyes the flowers look somewhat like insects, and apparently they look even more so to the male bees, for they fly at the flowers and butt them hard. The pollen becomes attached to the bee's head and is passed on to the next flower he butts.

Stranger still and as yet unproven is the possibility that some orchids mimic the prey of certain solitary wasps. Many kinds of wasps do not live in colonies. The female hunts caterpillars or other insects and stings them, injecting a venom which paralyzes but does not kill. She then carries the larva back to a nest she made in the ground and lays eggs on it. She closes up the nest and leaves. When the larvas hatch, they have a fresh supply of food which is enough to last until they are ready to form pupas.

One scientist observed a solitary wasp making stinging movements on the hairy lip of one kind of orchid. The orchid resembled a hairy caterpillar, and the wasp was a kind that lays eggs in caterpillars. Perhaps someone will study this strange interaction of flower and insect to see if the wasp is really "fooled" into thinking the flower is a caterpillar.

8 · Puzzles, Problems, and Proof

Many books this size could be filled with examples of mimicry, with at least one volume about mimicry in beetles alone. There are whole groups of mimicries with beetle models basically similar to those of tropical butterflies. Often these groupings include moths and grasshoppers which "copy" distasteful beetles. Unusual kinds of mimicry exist in beetles as well. One perfectly harmless beetle copies an unpleasant relative. The model does a sort of "headstand," raising its abdomen high in the air and releasing a drop of smelly, irritating fluid from the abdomen tip. The mimic has a shiny black body like its model and does the same headstand routine when disturbed, but it completely lacks the defensive secretion. Predators such as lizards and skunks are taken in by the act and give the mimic a clear berth.

Interesting "tidbit" examples of possible mimicry can be found in a vast array of animal groups. A Brazilian caterpillar, for instance, looks for all the world like a big, hairy spider. Its body has eight leglike lobes on the sides and it sits freely exposed on top of a leaf. Mimics and

An Eleodes bombardier beetle does its "headstand" before spraying an irritating secretion at predators. A mimic that looks like it does the same headstand but has no secretion; nevertheless, experienced lizards and other predators take the hint and go elsewhere.

models can be from completely unrelated animal groups. A small amphipod (a relative of shrimps and crabs) which lives among marine eelgrass appears to mimic snails which also live on the eelgrass. While most amphipods swim frequently, this kind rarely swims. It occurs in several color patterns, some dark, some light, and some with bands of brown and white. Snails with all these patterns can be found. The amphipod has a fat body, and when it crawls along the eelgrass, it moves with the same kind of rocking motion as do the snails. The outward resemblance of these two very different animals is remarkable, and apparently predators confuse them. Fish rarely eat the snails, probably because of their hard shells. In aquarium tests, too, they left the mimetic amphipods alone. Only if an amphipod betrayed its disguise by swimming did it get eaten. Another kind of amphipod which readily swam was quickly eaten by the same fish which left the mimics alone.

In the Kalahari desert of southern Africa lives a pale reddish-tan lizard which blends in well with the sand. Close relatives of this lizard often have young which are different in color from the adults. But this kind has young so completely different in coloration and behavior that it is hard to believe the two are related. The young lizards are jet-black with irregular whitish stripes on their backs and sides. Their tails are light in color. Unlike their camouflaged parents, the young lizards stand out against the pale sand. While the adults move with a typical lizard gait, waving their bodies from side to side as they go, the young move with a very peculiar stiff, jerky motion, their backs arched up and their tails pressed against the sand. Such obvious and slowly moving animals should be easy targets

for hungry birds, foxes, and snakes. But scientists studying them found that very few of them had the broken tails
which are telltale signs of predator attacks on lizards.
How, then, are they protected?

Scientists working in the Kalahari have sometimes seen
what they thought was an oogpister beetle marching
across the sand and discovered on closer inspection that
it was really one of these young lizards. The oogpister
beetle is a common and obnoxious fellow. When disturbed,
it squirts a very unpleasant acid secretion, so it is avoided
by possible predators. The oogpisters are black with white
markings, like the young lizards. They move with a
similar gait and are the same size. When the young lizards
get to about the maximum beetle size, they change their
color and their way of moving.

"Proving" Mimicry

While evidence that the young lizards could be mistaken for oogpister beetles is quite strong, there is always
the question of how to "prove" that mimicry is operating.
The kinds of predators involved in such a mimicry situation
are hard to work with in captivity, and duplicating the
desert environment in the laboratory would not be easy.
Observations in nature are difficult, too, for scientists
would have to run across beetles and lizards in the desert
and hope to see them interact with predators. Since most
animals do not behave normally when human observers
are near, such field studies would be just about impossible.
So mimicry can be judged to be only "probable" on the
basis of several facts: the beetles are conspicuous and ob-

noxious; the lizards copy them in size, coloration, and gait; the two animals live in the same places; and the beetles are much more common than the lizards. Also, the evidence that these lizards have a very low frequency of broken tails could indicate that predators do not often attack them. On the other hand, it could mean that the lizards' clumsy gait does not allow them to escape easily if they are attacked.

Some other proposed mimicry possibilities are even harder to prove or disprove. One scientist suggests that young cheetahs are mimics of the vicious and tough-skinned honey badger of Africa. Both the cheetah kittens and the honey badger are light on top and dark below, making them conspicuous. The young cheetahs have an almost white or silver-gray back, with dark fur below. No other member of the cat family has such a pattern, but it is quite similar to the coat of the honey badger. The cheetah kittens lose their young pattern when they are about two and a half months old, about the time they become bigger than the honey badger. Honey badgers are hardly ever attacked by predators, for their tough skin, sharp teeth and claws, and unpleasant odor protect them well. Cheetah kittens, however, are preyed upon by such powerful predators as lions, hyenas, and leopards. If they could be mistaken at times for honey badgers, it would certainly be to their advantage.

But what kind of evidence could a scientist present to "prove" mimicry in a case like this? Laboratory experiments are clearly out of the question, and field observations are even more difficult with large mammals than with lizards and beetles. We will probably have to be content with an interesting possibility, a believable variation on the theme of mimicry which is so frequently encountered in smaller, more abundant animals.

How Do Animals See Things?

When studying the lives of other creatures, we are always limited by our human perceptions of the world. Our eyes are very keen and can see many colors. Insect eyes are very different from ours in structure, and the range of colors they see is not the same as ours. We can never know for sure what an insect sees when it looks at something. And because our brains are different from those of other creatures, we cannot know what impressions their senses make upon them. Our sense of smell has such a minor role in our lives that it is hard for us to conceive of the importance odor has to other animals and to evaluate potential examples of odor mimicry. We cannot hear ultrasonic sounds, but bats can. Some distasteful moths emit ultrasonic clicks which bats can hear, and at least one edible moth does the same. Are the clicks warning sounds, functioning like warning colors? And is the edible moth a sound mimic? How can we study this problem when bats do not fly well in the enclosed laboratory spaces necessary for human investigation?

We can never fully penetrate the perceptual world of other creatures, and this limits our ability to decide what is mimicry and what is not. Many hole-nesting birds such as chickadees and titmice make a hissing sound if they are disturbed while incubating eggs. To us, the hiss resembles the sound of a snake, but does a potential predator perceive it in this way, or is the hiss simply a warning that the bird will fight to protect its eggs? We have no way of knowing.

An Eye for an Eye?

This problem of perception becomes especially important when we study one very common protective device of animals, false eyes. Fish, moths, butterflies, grasshoppers, peacocks, and various other animals have some sort of eyespot markings. Deciding what function eyespots serve is no easy matter. We call them eyespots because they remind us of eyes; that does not mean that other animals perceive them that way. Each type of eyespot situation must be studied on its own to determine its function. In some, such as the peacock, eyespots are involved in communication within the species. In others, the spots seem to startle possible predators. One ridiculous kind of South American frog has eyespots on its rear end. When a predator threatens, the frog lowers its head and presents its rear to its enemy, presumably scaring it off. But even in these cases, we cannot know if the predator is merely startled by the striking pattern presented or if it is actually frightened because it is "fooled" into thinking that the spots are the eyes of a potential enemy.

A number of years ago, Dr. A. D. Blest experimented with eyespot patterns in butterflies and their effects on birds. He tried several different approaches to the problem. He fed captive birds mealworms and projected different kinds of patterns next to the worms just as the birds were about to peck. He found that the birds were much

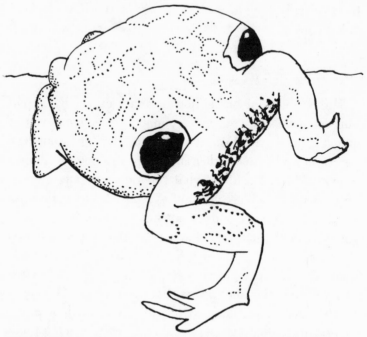

This South American frog, when disturbed, drops its head and raises its rear end with its two impressive black eyespots. The dark colors along its thighs may also look somewhat like a mouth to predators.

less startled by crosses or parallel bars than they were by circular patterns. They were most shocked by a pattern which resembled focused vertebrate eyes. Perhaps they did indeed react because circular patterns are reminiscent of eyes.

Blest also tested the reactions of birds to the eyespot pattern of the peacock butterfly. When resting, this butterfly holds its wings closed. The undersides have a brown camouflage pattern which blends in with the bark of trees. But if the butterfly is disturbed, it opens its wings, exposing two pairs of eyespots, one on each set of wings. As it opens the wings, a hissing sound is produced by the wing veins rubbing together. The insect can repeat the pattern of opening and closing its wings several times until danger passes.

When Blest tested the butterflies with birds, he found that inexperienced birds flew away when the butterflies put on their display. If a butterfly was presented over and over again, however, the bird would lose its fear and not leave. If Blest rubbed the colored scales off the wings, eliminating the eyespot pattern, the butterfly display failed to scare off birds, and the butterflies were eaten. This is pretty strong evidence that the eyespots on butterfly wings do serve to alarm potential predators.

But we must be careful not to generalize too much from one study, for some eyespots appear to have a very different function. Many butterflies have very small eyespots around the edges of their wings, and some wasps and beetles have them near the sting, or near glands which spray defensive secretions. Such small "eyes" seem to have the opposite effect from the large ones; they actually attract the pecking attacks of birds. If a bird attacks a butterfly with small eyespots around the wing edges, it is likely to peck at the spots rather than at the body of the insect. The butterfly might then be able to escape, with only slightly tattered wings to show for its bad experience. A wasp with small eyespots near its sting would be able to attack a predator on its vulnerable head if pecked there, and a beetle which was pecked near its defensive glands could spray a bird right in the eyes.

Some caterpillars have eyespots, too, and they may well have different effects, depending on the particular species. Swallowtail caterpillars have a gland just behind the head which produces a strong-smelling defensive secretion. Some kinds of swallowtail caterpillars also have a pair of false eyes right near this gland. Presumably this allows them to direct the attack of birds right to the place where they will get the strongest whiff. Other caterpillars have eyespots near their tail ends which would direct the attack towards a less vulnerable part of the body.

Hawkmoth caterpillars have yet another type of false eyes. These eyes are associated with other markings which give the overall impression of a snake's head. Not only eyes, but reptile-like scale patterns and a false "mouth" may be there. One kind even has what looks like a forked snake tongue. A tropical American hawkmoth caterpillar lives on vines at the forest's edge. From a distance it is camouflaged. But if its vine is jiggled, the caterpillar lets go with its front legs, swells up its body to reveal a pair of very realistic eyespots which were hidden before, and sways back and forth like a snake. The whole act makes the harmless caterpillar look very much like a poisonous viper about to strike.

Most butterflies with eyespots on their wings have the spots on the upper side of the wings, where they are hidden from view unless the butterfly is flying or has been disturbed. The butterfly is camouflaged at rest. But in the American tropics there are a few species which have the eyespots on the underside of the wings. These are among the largest butterflies in the world. One kind has huge

The front end of a spicebush swallowtail caterpillar, with the help of eyespots and a dark mark suggesting a mouth, resembles a small snake called the rough greensnake. The eyespots are the solid dark ovals with circles around them; the actual eyes are at the front of the head, which is turned down.

eyespot markings, 15 to 20 millimeters (over half an inch) in diameter. Another kind has much smaller spots, only six or seven millimeters across. The rest of the markings on the wings are light brown and gray, in streaks and patterns which blend in with the bark of trees where the butterflies rest during the day. From a distance the eyespots are not especially noticeable, but are striking from close up.

A scientist who was puzzled by these butterflies took a closer look at them, trying to figure out why these species, unlike other butterflies with large eyespot patterns, should always have the eyespots exposed. He noticed that the area around the large eyespot on the one kind of butterfly was different from the rest of the wing. It was dark, and had another, smaller circular marking in front of it. The whole darker area was set off by a light border, making it appear separate. The scientist finally concluded that the whole area around the big eyespot looked very much like the head of a large treefrog. The shape of the dark area was similar to the shape of a treefrog's head, and the smaller circle in front of the eyespot looked remarkably like the eardrum of a treefrog.

When he examined the butterfly with the smaller spots, he decided that its eyespot and surrounding area looked like the head of an anole lizard which lives in the trees. Since anoles feed on butterflies, he felt that both the tree-frog-mimic pattern and the lizard-mimic pattern were directed at the lizards. Those anoles are territorial and avoid one another. Larger lizards tend to win out in fights over territories, and the "lizard" on the butterfly wing is somewhat larger than a typical anole. Presumably then, a hungry lizard would be fooled into thinking that the butterfly wing was another larger lizard and would go away instead of attacking. But what about the treefrog pattern? The treefrogs apparently do not eat butterflies. But they are large enough to eat the lizards, so the treefrog-mimic pattern should look like a potential predator to the lizard.

No experiments have been done yet to see just how the lizards actually do react to the butterfly-wing patterns, so these speculations are only interesting ideas which need to be tested, just like so many other interesting potential examples of mimicry. The possibilities for mimicry are enormous, but the proven facts are very few outside of butterfly studies. Once scientists begin to concentrate effort on experimenting with some of these other fascinating mimicry situations, still more exciting information will unfold.

Glossary

abdomen: The usually long rear part of an insect's body

aggressive mimicry: Mimicry in which a predator or parasite "copies" something attractive or harmless in order to approach prey

Batesian mimicry: Mimicry in which a harmless species resembles a dangerous or distasteful one

cardiac glycoside: A poisonous chemical found in danaid butterflies, which they obtain from milkweed plants as caterpillars

danaid: A butterfly belonging to the family Danaidae, which includes monarchs and queens. Danaids are generally distasteful and warningly colored.

heliconid: A butterfly belonging to the family Heliconidae, the passionflower butterflies, which have narrow wings and which taste bad

Mertensian mimicry: A proposed type of mimicry (it may not exist) in which the models are moderately poisonous and the mimics are deadly poisonous; proposed by Wickler to apply to mimicry in coral snakes

mimic: An organism which "copies" the appearance, sound, odor, or behavior of another well enough to "fool" some particular observer

mimicry ring: A group of three or more species which resemble one another

model: The organism which a mimic "copies"

Müllerian mimicry: Mimicry in which two or more distasteful species resemble one another

pierid: A butterfly belonging to the family Pieridae, which includes the white cabbage butterfly and the sulphur butterflies

thorax: The middle portion of an insect's body. The wings and legs attach to it.

Suggested Reading

Books

Michael Fogden and Patricia Fogden, *Animals and Their Colors: Camouflage, Warning Coloration, Courtship and Territorial Display, Mimicry* (Crown, N.Y., 1974). An adult book full of interesting information and fine photographs.

Dorothy Hinshaw Patent, *Plants and Insects Together* (Holiday House, N.Y., 1976)

Dorothy Shuttlesworth, *Animal Camouflage* (Natural History Press, N.Y., 1966). Mostly about camouflage rather than mimicry.

Hilda Simon, *Insect Masquerades* (Viking Press, N.Y., 1968)

Otto von Frisch, *Animal Camouflage* (Franklin Watts, N.Y., 1973)

Wolfgang Wickler, *Mimicry in Plants and Animals* (World University Library, London, 1968). An adult book with many examples and fine illustrations, written by an expert.

Magazine Articles

Lincoln P. Brower, "Ecological Chemistry," *Scientific American*, Feb. 1969. About blue jays and monarchs.

Lincoln P. Brower and Jane v Z. Brower, "Investigations into Mimicry," *Natural History,* April 1962

H. Hediger, "Camouflaged Still-fishing," *Natural History,* June 1963. Worm-lure of the alligator snapping turtle.

Bert Hölldobler, "Communication Between Ants and Their Guests," *Scientific American,* March 1971

Nicol Niaolai, "Mimicry in Parasitic Birds: Widow Birds," *Scientific American,* Oct. 1974

Patricia Raymer, "Something's Fishy About That Fin," *National Geographic,* Aug. 1974

George F. Rohrmann, "Misleading Mantids," *Natural History,* March 1977

Edward S. Ross, "Asian Insects in Disguise," *National Geographic,* Sept. 1965

Miriam Rothschild, "Mimicry: The Deceptive Way of Life," *Natural History,* Feb. 1967

John R. G. Turner, "A Tale of Two Butterflies," *Natural History,* Feb. 1975. About mimicry in the passionflower butterflies, with color photos.

J. Welsh, "Mussels on the Move," *Natural History,* May 1969

Index

African mocker swallowtail, 31–35

aggressive mimicry, 13–14, 15, 44–61, 84, 94

amphibians, 62; see also frog, salamander, toad

amphipod, 107

anglerfish, 45–48; deep-sea, 46–47

antennas, 80, 81, 84

ants, 7, 18–19, 53, 54, 81–84; army, 87; larvas, 88–89; weaver, 82

aphids, 53, 85; woolly, 53–54

barley, 97

Bates, Henry Walter, 8–10, 34

Batesian mimicry, 16, 17–18, 25, 31, 35, 71, 75, 78, 94

bats, 99, 110

bees, 7, 14, 93–95; bumblebees, 94, 95; cuckoo, 94–95; honeybees, 93–94

beetles, 7, 18–19, 51, 85–89, 105, 113–114; bombardier, 7, 106; ladybird, 18; oogpister, 108–109; tiger, 7, 13; see also fireflies

behavior, 24–26, 50–51, 54, 64–66, 76, 77–78, 87, 89, 93–94

berries, mimics of, 98–99

birds, 62, 110; as predators, 12–13, 15, 20–23, 39, 72, 81, 84, 85, 111–113, 114; cuckoo, 13, 56–57; eggs, 56–57, 59; food choice of, 25–26; gape patterns, 59–66; how eats a butterfly, 24–26; indigo, 57–61; mimicry in, 13, 56–61; songs, 57, 61; see also specific kinds

Blest, A. D., 111–113

blue jay, 17, 21–22, 23, 24–26

brood parasites, 56–61

Brower, Jane and Lincoln, 21–28, 94

bugs, 7

butterflies, 8–13, 15, 20–43, 111, 114–117; admiral, 30–31, 33; African queen, 31; cabbage, 8; females as mimics, 27, 34–35; friar, 30–31, 33; mating behavior of, 34–35, 41–42; passionflower, 8–11, 36–42; peacock, 112–113; queen, 28; viceroy, 20–21, 23, 28; see also (cont'd)

African mocker swallowtail,
Danaidae, monarch butterfly,
Pieridae

camouflage, 14–15, 15, 42, 43,
44, 66, 74, 85
cardiac glycosides, 23–24; distribution in monarch body,
24; potency of, 24
caterpillar, 6, 16, 37, 84, 105,
114; of hawkmoth, 114; of
monarch, 21–22; of swallowtail, 114, 115
cheetah, 109
*Coatonachthodes ovambolandi-
cus*, 92
Colloplesiops altivelis, 68–70
crickets, 7

Danaidae, 30–31
dandelion, 97
Ditmars, Raymond L., 49
ducks, black-headed, 59
Dutchman's pipe, 101

evolution, 11–14, 16–18, 39–42,
56–57, 60–61, 73, 76–78
eyespots, 6, 111–117

fireflies, 50–51
fish, 54, 62–70, 111; as predators, 108; batfish, 67–68;
blennies, 63, 64, 65–66;
cleanerfish, 64–66; damselfish, 63; learning in, 65–66;
mimicry in, 13–14, 62–70;
moray eel, 68–70; orange anthiid, 64; reef, 61–70; sea
swallow, 64–66; *see also*
anglerfish, scorpionfish
flatworm, 66–68

flax, 14
flies, 7, 99–102; bee, 9; droneflies, 93–94; hoverfly, 94; narcissus bulb, 94; robberflies,
94
frog, 112, 116

glochidia, 54, 56
gold-of-pleasure, 14
grasses, 99
grasshoppers, 7, 13, 105, 111
great kiskadee, 77–78
green lacewing larva, 53–54
grooming, 88

Heliconidae, *see* butterflies, passionflower
Heliconius erato and *H. melpomene*, 37–42
heredity, 8, 11–13, 76–77, 78
honey badger, 109
honey guide, 57–58
hornets, 7
humans, 23–24
hummingbird, 82, 99

Ice Age, 39, 40
insects, 25; social, 79–92; *see
also specific kinds*

Kistner, David, 92

lizards, 50, 105, 116
lure, fishlike, 47, 48, 49, 55; of
mussel, 55; snake tail, 48–50

maggots, 93, 101
mammals, 62
mantis, 85; flower, 45; Malaysian flower, 44; praying, 44

Mertensian mimicry, 75, 78
milkweed, 21, 22
mimic, 7, 12, 17, 72
mimicry rings, 35–36, 42–43, 73–74
model, 7, 12, 17, 72, 75, 76
monarch butterfly, 17, 20–28, 30; California, 26, 27; Massachusetts, 22, 27; Mexico, 26; migration of, 26–28; Trinidad, 22, 28
moth, 7, 36, 52, 105, 110, 111; cinnabar, 16; hawkmoth, 114; lobster, 84–85; sphinx, 6, 15; tiger, 18
Müller, Fritz, 10
Müllerian mimicry, 16, 35, 37, 42, 78
mussel, 54–56

nectar, 99

oats, 97
odor mimicry, 18–19, 86, 88, 90–91, 110
onion, wild, 98
orchids, 14, 101–104; Andean, 102; ophrys, 102

Papageorgis, Christine, 42–43
parasites, 54–56; brood, 56–61, 95
peacock, 111
pedipalps, 80
petiole, 87
Pieridae, 8–11
pines, 99
pit vipers, 49
plants, 14, 96–104; pollination of, 99–104
poisons, 21, 23–24, 26–28, 70–71; *see also* cardiac glycosides; snakes, poisonous
predators, 16–18, 19, 35, 40, 43, 44, 75, 78, 81–82, 93–94, 105, 108; *see also* aggressive mimicry; birds, as predators; fish, as predators

red eft, 70–73
reptiles, 62; *see also* lizards, snakes
rye, 96–97

salamander, red, 71
scorpionfish, 47–48
seeds, 96, 97; fruitlike, 97–99
sex attractants, 52
skunks, 105
Smith, Susan, 77–78
snails, 74, 107
snake, 48–50, 110, 114, 115; banded rock rattler, 50; cobra, 73; colubrids, 73, 76; copperhead, 50; coral, 73–78; cottonmouth, 50; indigo, 19; krait, 73; mamba, 73; poisonous, 48–50, 73–78; Saharan sand viper, 50
sound mimicry, 19, 110
species, 11
spiders, 80–84; bolas, 51–53; crab, 14–15; jumping, 82; wolf, 83
starfish flower, 101

termites, 18, 89–92
toad, 93–94
touch mimicry, 18–19, 91
Turner, John R. G., 39–42

Virgil, 93

wake-robin, 100
warning colors, 15–16, 25, 28,
    36, 42, 73–74
wasps, 7, 16, 82, 103–104, 113–
    114

wheat, 96–97
Wickler, Wolfgang, 75
widowbirds, paradise, 59; *see
    also* birds, indigo
woodpeckers, 59